The SOUTHERN ITA FARMER'S TAB

AUTHENTIC RECIPES AND LOCAL LORE FROM TUSC

Matthew Scialabba & *Melissa Pellegrino*

AUTHORS OF *THE ITALIAN FARMER'S TABLE*

AN IMPRINT OF GLOBE PEQUOT PRESS
GUILFORD, CONNECTICUT

To Japhy and her insatiable appetite.

To buy books in quantity for corporate use
or incentives, call **(800) 962-0973**
or e-mail **premiums@GlobePequot.com**.

©2012 Melissa Pellegrino and Matthew Scialabba

ALL RIGHTS RESERVED. No part of this book may be reproduced or transmitted in any form by any means, electronic or mechanical, including photocopying and recording, or by any information storage and retrieval system, except as may be expressly permitted in writing from the publisher. Requests for permission should be addressed to Globe Pequot Press, Attn: Rights and Permissions Department, P.O. Box 480, Guilford, CT 06437.

Lyons Press is an imprint of Globe Pequot Press.

Project editor: David Legere
Text design: Sheryl Kober
Layout artist: Nancy Freeborn

Photos by Melissa Pellegrino and Matthew Scialabba unless otherwise noted.

Library of Congress Cataloging-in-Publication Data is available on file.

ISBN 978-0-7627-7082-3

Printed in China

10 9 8 7 6 5 4 3 2 1

CONTENTS

ACKNOWLEDGMENTS

Five months of schlepping across central and southern Italy requires stamina and endurance and a somewhat organized itinerary—which wasn't always the case for us. From our home in Connecticut, using the Internet, books, and personal recommendations, we weeded through the thousands of agriturismi out there, hoping to select a variety of different farms based on their agricultural activities, style of cooking, and accommodations. An e-mail was sent, claiming our intentions and asking the farms if they would be willing to have two Americans enter their lives for a few days, to learn and absorb their way of life by spending time in their fields and kitchens. The responses trickled in slowly; farmers aren't always checking their inbox. This book would not have been possible without the generosity and enthusiasm of each agriturismo that welcomed us with open arms and brought us into their lives and homes. *Grazie mille* to everyone at Sa Mandra, Sa Tiria, Muto di Gallura, Giardino di Vigliano, Il Cortile, Porta Sirena, Santa Marina, Dattilo, Le Puzelle, Carrera della Regina, Torrevecchia, Narducci, Serragambetta, Masseria Santa Lucia, I Dolci Grappoli, Pietrantica, Le Magnolie, Campoletizia, Costa della Figura, Giardino degli Ulivi, Villa Dama, Malvarina, Fonte Antica, Terra Etrusca, San Martino, Casale Verdeluna, Mole sul Farfa, Santa Mamma, Tenuta Roccadia, and Baglio Fontana.

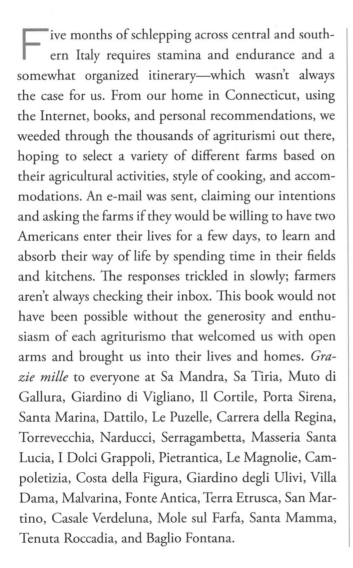 Once again to our families, for their unconditional support.

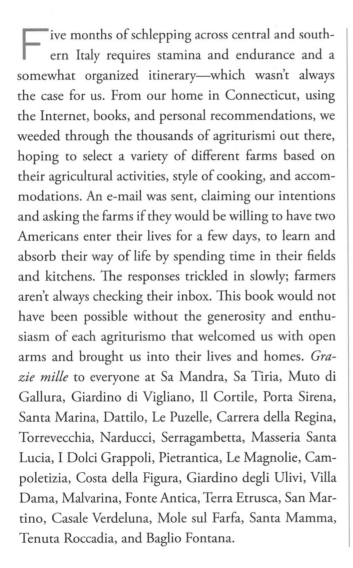 To our great friends we are lucky to have.

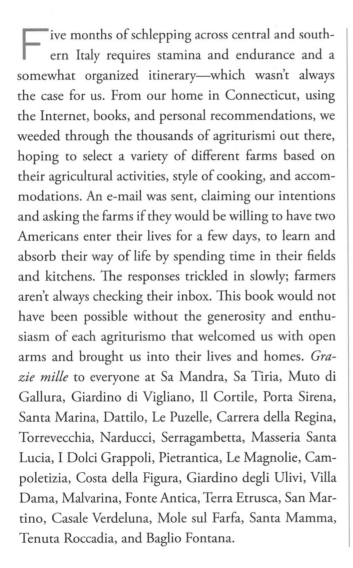 To Deb and Jim, for your culinary inspiration.

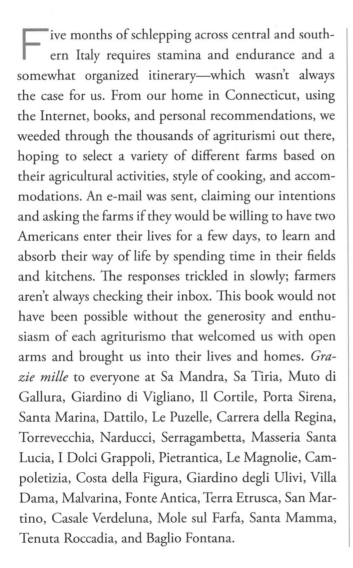 To Liana and Tony, for teaching us about your southern Italian roots.

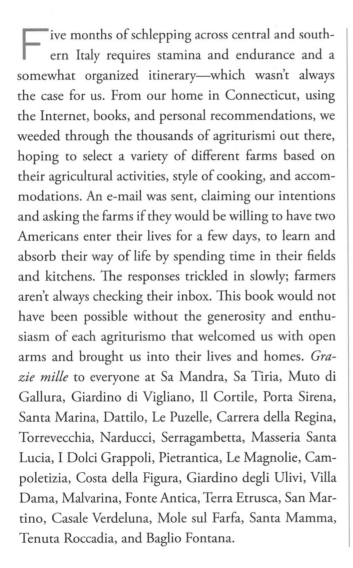 To everyone who bought our first book that made this sequel possible.

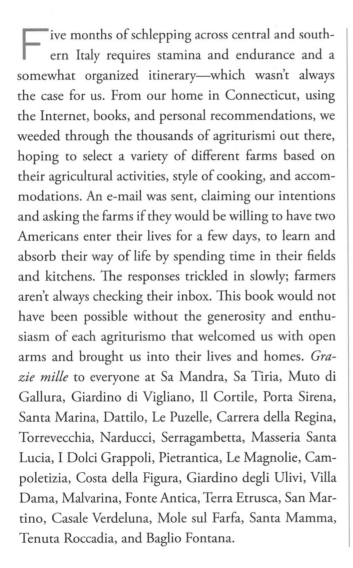 To Bishop's, Forte's, and Star Fish for making shopping in Guilford feel like we are back in Italy.

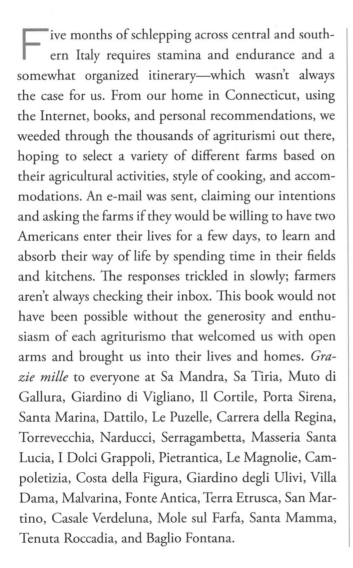 To Mary Norris and everyone at GPP for making this dream happen.

INTRODUCTION

The Italian agriturismo system was developed to preserve the act of farm-to-table eating long before it became a trendy slogan. It was born from the need to help the thousands of family farms across the country that were struggling to survive—yet essential lifelines to the pulse of Italian cuisine. Opening their doors to overnight guests and diners allowed farmers to supplement their incomes and showcase the flavors of their land by providing meals prepared with their own seasonal ingredients. The success of the movement penetrated deep into rural Italy, as abandoned farms were restored to their former brilliance, and travelers from around the world embraced the concept of a vacation soaked in the splendors of the Italian countryside. Today the Italian agriturismo symbolizes a revival of times past; it speaks of practical healthy living that is sustainable and a model of living off the land.

After our first experiences working on several Italian agriturismi upon graduating from culinary school, we were deeply moved by the simplicity of cooking with ingredients grown and raised out the kitchen door. There was something that we connected with, that just seemed real and felt right about washing dirt off freshly picked vegetables, or noting the vibrant orange color of the free-range egg yolks we were using to make fresh pasta. We grew enamored with the magic that happened in cooking with an entire farm at our disposal. We returned home with this philosophy engraved in our budding culinary minds, and hatched a plan that was full of unbridled youthful dreams—to write a cookbook about the lives and recipes from agriturismi throughout all twenty regions of Italy. Through tenacity, determination, and a little luck, we completed half of our goal with our first cookbook about the agriturismi of northern Italy, *The Italian Farmer's Table*. And now, almost a decade after our initial agriturismo inspiration, down the tortuous road of life's twists and turns, we bring to you the sequel book, which completes this chapter of our lives.

Our first book ended in Emilia-Romagna. Our immersion into the core of the central and southern regions thus began in the hills of Tuscany, down the less traveled roads of the deep south, out to the faraway land of Sardegna, and finally ending on the shores of Sicily. With the incredible journey and experiences of our first trip still fresh in our minds, initial comparisons to our northern voyage were difficult to avoid. But as we eased into our routine and the trip began to develop its own unique character sketched from the stories of the people, places, and food we encountered along the way, a revelation of Italy's beauty came to us. Perhaps its most apparent charm lies within the diversity found in each region, city, and village, which is most obvious in the immense chasms dividing the north, central, and south. Within the confines of an entire country not much larger than the state of Arizona, there exist vast differences in landscape, climate, terroir, and personalities. These shape the food cultures of each region that are so fiercely loved and protected by the populace. There is no better way to experience this great diversity than by visiting the country's agriturismi, packed with local flavor and color. Their backbones are the heirloom recipes prepared with each farm's own products, which speak of the true essence of Italian cuisine. And that's exactly what we did.

During our five-month culinary adventure at over thirty of these farms, we unearthed and discovered Italy's rural soul. Each day brought us new inspiration and moments of awe at the marvels and depths of Italy's gastronomy. We arrived just in time to help with the saffron harvest, and plucked its fire-red thread, staining our fingers with a burnt orange hue. We visited countless cheese farms, including numerous pecorino producers in the central hills that are home to flocks of grazing sheep; a Sardinian dairy that had us sample the mouth-singeing, eye-watering island delicacy of maggot-infested cheese; and a stone hut in Sicily, where we witnessed the dying tradition of making cheese in a copper cauldron over a fire. Its taste leaves a lingering hint of smoke on the palate. Our pasta lessons were numerous. At each farm we were exposed to countless shapes and types, clouding our memories with the innumerable variety in the Italian repertoire. After spending a morning watching the crooked and bent fingers of a ninety-seven-year-old

grandmother roll and cut perfect little ears of orecchiette, we drank to her health with rounds of limoncello. We spent time among the ancient olive groves of the Sabina park of Lazio, whose twisted, knobby trunks bear spine-tingling evidence of Roman origins, and drizzled their liquid gold over everything we ate there. We helped net fat healthy fish at a visit to a trout farm fed from a mountain stream, picked grapes for the 2009 vintage at an award-winning winery, and went looking for mushrooms in damp Calabrian forests. A farm in the heart of water buffalo country had us falling in love with the gentle beasts before even taking a bite of the insanely delicious mozzarella. Some farms put us right to work eager to take advantage of our free labor, and we waited tables, our American-accented Italian revealing our nationality to diners, and cooked on the line, feeling the pulse of the bustling agriturismo kitchen.

Five months traversing Italian roads was not without excitement. A shredded tire and smashed bumper bear testament to our calamities. The crater-size pockmarked roads of the deep south were often blocked with protestors or herds of animals, opening our eyes to the sorry state of the infrastructure of the poorest regions. In the mountains of the Abruzzo, we listened to the stories of a retired shepherd, who recounted Italy's past of world wars and poverty through colorful narrations about his simple life of tending to his animals. After witnessing the slaughter and butchering of suckling pigs, we politely accepted a pig's ear cut from one of the animals after it had been passed over with a blowtorch. It was uniquely chewy. An

Umbrian farm cooked us a local specialty of grilled air-dried pigs' intestines—decidedly not our favorite—and a visit to an antique fountain to fill water bottles proved to be an ill-fated decision resulting in sleepless nights, a longing for home, and a visit to an Italian hospital. We begged and pleaded for a family's highly coveted award-winning recipe—and finally prevailed after winning over the mom with our charm. We helped pick olives, joining in the camaraderie of the fall activity, and were invited to harvest feasts at long wooden tables festive with banter fueled from the invigorating work and jugs of wine. The explosion of sweet citrus flavor from blood oranges and clementines picked straight from the tree made us gasp at the intensity, leading to severe stomachaches from overindulging. Lunch with a colorful set of characters at a hunting lodge deep in the forests of Tuscany showed us the finer intricacies involved in dismembering a wild boar, and a morning at a local horse show revealed the grace and beauty of the region's working horses. The generous spirit of the south quickly taught us that saying no to a second helping is impossible. The same goes for thirds. Foraging for wild greens illustrated the thriftiness of Italians who can make something delicious from a field of weeds, and baking bread in a brick oven and grilling thick porterhouse steaks over smoldering coals proved to us the unrivaled goodness of cooking with wood.

All these experiences were eye-opening, yet the true beauty behind them is that they are simple, ordinary happenings that occur every day in the food world of Italy. The undeniable fact is that Italians simply know how to eat. They can talk about what is for dinner while eating lunch, and why a dish from their own province is so much better than that of their neighbor when the two are virtually identical. A common thread throughout the country is that quality of ingredients is crucial, including understanding where they are grown and how they are made. And that is why agriturismi are so special. We hope that the stories and recipes on these pages offer a taste and flavor of Italian working farms, and inspire you with the purest examples of the Italian philosophy so fundamental to its cuisine.

BASIC RECIPES

Basic Semolina Pasta Dough

3 cups (375 g) semolina flour
½ teaspoon kosher salt
1–1¼ cups warm water

On a clean work surface, mix together the semolina flour and the salt. Make a well in the flour and add the water to the center of the well a little at a time. With a fork, gradually pull some of the flour into the water mixture, and continue mixing until a soft dough begins to form. With your hands, knead the dough until it is smooth and elastic, about 8 minutes. Cover the dough with a kitchen towel and let rest for 20 minutes. Cut the dough into desired shapes.

SERVES 4

Basic Egg Pasta Dough

PASTA FRESCA ALL'UOVO

2¼–2½ cups (280–300 g)
   all-purpose flour
4 large eggs, room temperature

1. Make a well in the flour and add the eggs to the center. With a fork, lightly beat the eggs, and use the fork to gradually pull in some of the flour. Mix the flour and eggs together until well combined. Knead dough until it becomes smooth and elastic. Cover the dough and let rest for 20 minutes.

2. Cut the dough into six pieces and cover with a towel. Pass one piece at a time through a pasta machine. Starting on the widest setting, pass the dough through, then fold the dough like an envelope and pass it through the widest setting again. Repeat these steps until the dough is smooth and elastic, 8 to 10 times. Begin to lower the setting a notch on the pasta machine and pass the dough through once on each setting, finishing on the second to thinnest setting.

3. Lay the dough out on a flat surface sprinkled lightly with flour. Let the dough dry slightly; it should still be a little tacky and pliable or it will not cut properly.

4. Cut the dough into desired shapes: tagliatelle, taglierini, spaghetti, what have you. Or use the sheets to make stuffed pastas such as ravioli, lasagna, or cannelloni.

SERVES 4

Crepes
CRESPELLE

Place the flour, water, eggs, and salt in a blender and pulse to a smooth batter. Heat a 10-inch nonstick skillet over medium-high heat, brush the pan with a little melted butter, and ladle in enough batter to coat the bottom of the pan. You want to make the coating as thin as possible. Cook the crepe until it begins to dry out and set, about 1 minute, then use a spatula flip the crepe over and continue to cook another 30 seconds to 1 minute. Remove from the pan and lay flat on a dish. Continue cooking the crepes in this manner until the batter is finished.

MAKES 12 CREPES

1 cup (125 g) all-purpose flour
1 cup (240 ml) water or milk
3 large eggs
Pinch of salt
1 tablespoon unsalted butter

Tomato Sauce
SUGO DI POMODORO

2 tablespoons (30 ml) extra-virgin
    olive oil
1 clove garlic, smashed
1 28-ounce (793 g) can whole plum
    tomatoes
Kosher salt and freshly ground
    black pepper

Put the oil and garlic in a 4-quart pot over medium heat and cook until the oil is fragrant and the garlic is lightly golden, 2 to 3 minutes. Add the tomatoes and a pinch of salt and cook for 30 minutes. Whisk the sauce vigorously to break up the tomatoes and garlic, then season to taste with salt and pepper.

MAKES 2 1/2 CUPS

Béchamel
BESCIAMELLA

¼ cup (½ stick) unsalted butter
¼ cup (40 g) all-purpose flour
3 cups (720 ml) milk
Pinch of nutmeg
Freshly ground white pepper
Pinch of salt

1. Melt the butter in a 3-quart pan over medium heat. Add the flour and cook until the mixture is a golden brown, about 2 minutes.

2. Meanwhile, warm the milk in a 2-quart saucepan over medium-low heat. Whisk the milk into the butter mixture a ladleful at a time and bring up to a simmer. Cook until the sauce has thickened. Add the nutmeg, white pepper, and salt.

MAKES 3 CUPS

Chicken Broth
BRODO DI POLLO

1. Place all the ingredients in a large stockpot and cover with water. Bring to a slow boil. Reduce the heat and simmer gently for 2 to 3 hours, skimming any scum from the surface.

2. Strain the broth into a clean bowl and refrigerate until needed. Remove any hardened fat that rises to the top. Chicken broth can remain in the refrigerator for up to 5 days, or you can freeze it.

   MAKES 4 QUARTS

6 pounds (2.73 kg) chicken parts

2 yellow onions, cut into large dice

2 carrots, peeled, cut into large dice

2 stalks celery, cut into large dice

2 bay leaves

4 sprigs parsley

3 sprigs thyme

Vegetable Broth
BRODO DI VERDURE

Place all the ingredients into a large stockpot and cover with water. Slowly bring to a boil, reduce the heat, and simmer for 1 to 2 hours.

Strain the broth into a clean bowl and keep it in the refrigerator for up to 5 days, or freeze it.

MAKES 2 QUARTS

Other vegetables can be included in the broth, such as turnips, parsnips, asparagus, and red onion. Just steer clear of overly strong vegetables like cauliflower, cabbage, broccoli, and beets.

1 yellow onion, roughly chopped

2 leeks, cleaned and sliced

2 carrots, roughly chopped

2 stalks celery, roughly chopped

8 ounces (226 g) mushrooms, sliced

1 potato, peeled and cut into
   large dice*

1 bay leaf

1 sprig parsley

1 sprig thyme

1 teaspoon black peppercorns

Sponge Cake
PAN DI SPAGNA

Softened butter, for the pan

4 large eggs, separated

²/₃ cup granulated sugar, divided

1 teaspoon (5 ml) vanilla extract

1 cup (125 g) all-purpose flour

Pinch of table salt

1. Position a rack in the center of the oven and heat the oven to 350°F (180°C). Butter a standard 18 × 13-inch half sheet pan and line with parchment paper.

2. In a large bowl, beat the egg yolks, ⅓ cup of the sugar, the vanilla, and 2 tablespoons warm water with an electric mixer on high until thickened and very pale yellow, 5 to 6 minutes.

3. In a stand mixer fitted with a whisk, beat the egg whites on medium speed to soft peaks. Gradually pour in the remaining ⅓ cup sugar. Raise the speed to medium-high and continue beating until glossy stiff peaks form.

4. Fold a third of the egg whites into the yolk mixture to lighten, and then gently fold in the remaining whites. Sift the flour and salt over the top of the mixture and gently fold it in. Pour the batter into the prepared pan, gently spreading and smoothing it to make sure it's level. Bake until the top springs back lightly when touched, 10 to 12 minutes. Transfer to a wire rack and let cool completely.

I CAKE

The central regions of Sardegna, Tuscany, Lazio, Umbria, Le Marche, Abruzzo, and Molise are geographically and metaphorically the epicenter of Italy's gastronomy. They hosted two of the world's most influential and prosperous cultures, the Roman dynasty and Tuscany's Renaissance, both of which shaped and defined present-day Italy with their colossal achievements. This rich history has left Italians with inspiring architecture, masterpieces of art, and a cuisine based in the bounty grown and harvested from the same fertile soil as centuries past.

The recipes and traditional dishes of today's central Italy have influences from neighboring regions but are independently distinct, made from only local ingredients. Fresh pasta is prepared with both eggs and flour—as it's done throughout the north—as well as in the southern Italian manner with only flour and water. Bread eaten throughout Tuscany, Umbria, and Le Marche may strike the outsider as bland and tasteless from its lack of salt, but locals have a special place in their heart for it, and heavily season their cured pork products to accompany thickly cut slices. Leftover loaves are cubed and added to salads and soups.

Central Italy's aesthetically dramatic panorama of sloping hills and open meadow pastures that spread to the base of the rugged Appenines affords a landscape ideally suited to rearing sheep, and forged the shepherding cultures of the regions. Sardinian shepherds migrated to the mainland throughout the mid-1900s, bringing with them their pastoral lifestyles and their expertise at making sheep's cheese. Their assimilation and acquisitions of land and property made them permanent residents of their adopted regions, and cemented pecorino's reputation as the cheese of central Italy. Skewers of mutton have become synonymous with the Abruzzo and Molise, where tender cubes of the meat and fat are threaded onto sticks and grilled over wood coals. Their tenderness and earthy flavor are attributed to the animals' lives spent grazing in high-altitude plains.

Italy's most recognized steak, the infamous, insanely thick T-bone bistecca fiorentina, served bloody rare, comes from the Chianina cattle raised throughout Tuscany, Lazio, and Umbria. The stark white, well-muscled animals date back to Roman times, and were prized for their ability to work the steep hills of the regions. Tractors have now replaced the giant beasts in the fields, but a

Cork trees

smaller breed of Chianina is bred for its intensely flavored and marbled meat. Special black pigs with a white stripe known as Cinta Senese, also native to Tuscany, are now reared throughout other central regions; their succulent meat produces superior dried salamis and prosciuttos. Central Italy is also home to the country's most organized and skilled hunting clubs, which convene weekly throughout designated seasons to seek out a variety of the wild game that dominates menus from Sardegna to Molise. Wild boar, hare, pheasant, quail, and partridge all have their place in traditional country kitchens. Although the region is flanked by two oceans, fish plays a minimal role, even on the island of Sardegna; central Italians feel a closer connection to the land than the sea. Freshwater fish farms in Umbria and Le Marche produce fattened mountain trout fed from the natural rivers that slice through Sibillini National Park, and their fillets are a common second course among hearty meat fare.

Artichokes, peas, and fava beans are perennial favorites, as are heartier cold-weather vegetables like cardoons, Tuscan kale, and all sorts of squash. Seasoned foragers comb open meadows for an abundance of wild edibles, including chicories, bitter greens, and stalks of wild fennel. Dense forests damp with rainy fall and spring weather explode with spores of sprouting mushrooms, and beneath groves of poplar and beech trees specially trained dogs unearth knobby black truffles. Dark brown freshly tilled fields are seeded with wheat, lentils, beans, farro, spelt, bulgur, and corn in this cuisine based heavily upon legumes and grains. The concept of terroir—the perfect marriage of soil composition and climate—reaches a pinnacle in Tuscany's vineyards, and its wines are now regarded as some of the best around the world, but bottles of lesser-known varietals throughout the other regions should not be overlooked; they offer exceptional quality at more affordable prices. Olive oil is the ultimate unifier in every region. Tables are always set with bottles of locally produced extra-virgin oil to drizzle generously over each course.

The rustic good eating that has made Italian food famous around the world is alive and well in the belly of its central regions. Blessed with temperate weather and a fruitful landscape, central Italy has bestowed its grace and magnetic appeal to all those lucky enough to have spent time in its embrace and enjoyed its cuisine, one of the world's finest.

Resting bull

SARDEGNA (SARDINIA)

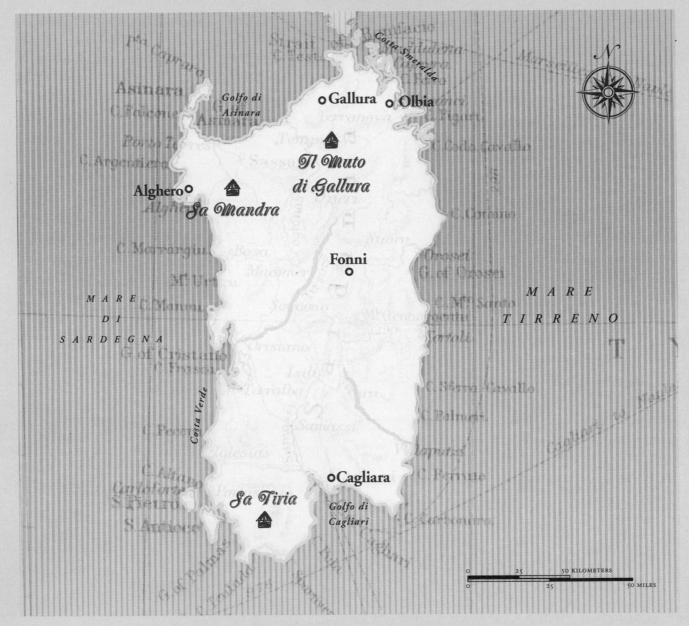

CHAPTER I

SARDEGNA (SARDINIA)

SA MANDRA

A true agriturismo in every sense, Sa Mandra is a model of farm-to-table eating. Husband and wife Mario and Rita and their children Maria Grazia, Michele, and Giuseppe have slowly built up over the years one of the most successful agriturismi in Sardegna. Mario and Rita both grew up in the mountainous hinterland region of Barbargia, which boasts the island's highest peak. The area's rugged climate of snow and ice is a far cry from the sandy beaches and clear blue waters of Sardegna's coast.

Both spouses were raised in large families of shepherds with strong roots to the cultural heritage of their land. They left home over twenty years ago, when life in the Barbargia had become unbearably difficult, as there were too many shepherds and not enough land to accommodate all of them. They moved to the coastal flatlands of the Alghero province, with the intentions of raising their family and starting a farm of their own, where they could continue the traditions of their families. They began slowly with a small plot of land and flock of sheep, naming it Sa Mandra, which refers to the pen where sheep are milked and is symbolic of their roots and identity. Soon after, they began offering meals in their home, with guests dining at their own kitchen table. Everything was prepared with ingredients from their land. In addition to meat, the family was also producing cheese from their sheep.

Much has changed since Sa Mandra's early years as one of Sardegna's first farm restaurants. Over the years the agriturismo has expanded to serve the growing numbers of tourists discovering the island's beauty. Today the farm hums with activity throughout the summer months. Small and wiry and with a look of eternal youth, Rita sparks the kitchen to life every night. Her daughter Maria Grazia, who passed up a law career to devote her life to her family's business, manages the dining rooms with grace and an impressive efficiency that sees hundreds of customers fed every night. She greets arriving guests

Mural depicting Sardinian culture

in the farm's courtyard with an apertivo of local wine and a selection of cheeses. They can stroll around the property, which is made up of small stone buildings that house miniature museums of antique tools and artifacts of rural Sardinian culture. Visitors are also encouraged to watch Mario in action when he cooks porcetto, the farm's signature dish, in enormous open fireplaces. The farm's own suckling pigs are skewered onto long metal spears and then set in front of a lively fire of burning olive and myrtle wood. They are rotated every hour for four hours until their skin becomes perfectly browned and crackling. Heaping portions are served to diners, who delight in the delicacy and texture of milk-fed pork with a hint of smoky flavor. Sa Mandra is famous across the island for their preparation of porcetto, and the fireplaces symbolize the farm's success and heritage.

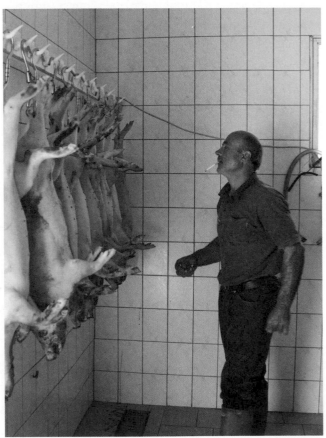

Just-butchered suckling pigs

Suckling pigs ready to be skewered

Porcetto, spit-roasted suckling pigs

Ravioli with Seven Herbs and Ricotta

RAVIOLI ALLE SETTE ERBE

1 ½ (680 g) pounds ricotta

¼ cup loosely packed basil leaves

¼ cup roughly chopped flat-leaf
  parsley

1 tablespoon roughly chopped mint

1 tablespoon sliced chives

1 tablespoon roughly chopped
  chervil

1 teaspoon roughly chopped
  oregano

1 teaspoon thyme leaves

Kosher salt

¼ cup (59 ml) extra-virgin olive oil

1 recipe egg pasta dough (see
  "Basic Recipes")

2 tablespoons freshly grated
  pecorino cheese, plus more
  for serving

At Sa Mandra, Rita simply walks out the kitchen door to her herb garden, picking what she needs to make the filling for these ravioli. They are packed with flavor, and the sauce made from a scoop of the filling adds a nice creaminess to the dish.

1. In a food processor fitted with the blade attachment, add the ricotta, the herbs, and 2 teaspoons of the salt. Process until the herbs are chopped. With the motor running, gradually add the oil in a steady stream until it is incorporated, the herbs are finely chopped, and the mixture is the consistency of mayonnaise. Reserve ⅓ cup of the cheese mixture and set aside.

2. Divide the dough into four pieces. Using a pasta machine, roll the dough out, starting with the widest setting and then going down until the dough is ⅛ inch thick. Lay out one sheet of pasta. Drop teaspoons of filling down one side of the pasta sheet at ½-inch intervals. Fold the pasta over the filling and gently press out any air between the filling and the dough. With a fluted pastry cutter, cut the ravioli into small squares (about 1½ inches) and place on a lightly floured baking sheet. Finish the remaining ravioli in the same fashion.

3. Bring a large pot of well-salted water to a boil. Drop the ravioli into the boiling water and cook until they rise to the surface, 2 to 3 minutes. Reserve ¼ cup of the pasta water, then drain the ravioli. Toss the ravioli in a bowl with the reserved cheese filling, a little of the pasta water, and the pecorino cheese. Serve immediately with more cheese if desired.

SERVES 5–6

Fregula with Lamb Ragù and Zucchini

FREGULA CON SUGO D'AGNELLO E ZUCCHINE

Sardinian fregula, sometimes referred to as Sardinian couscous, is eaten all over the island and made by mixing semolina flour with water. Traditionally water was sprinkled over the flour and then sifted until small, pebble-shaped clumps formed. The granules were then dried, toasted, and cooked in boiling water until al dente. Here the pasta is paired with a hand-cut lamb ragù. Taking the time to cut the meat into small cubes complements the small shape of the fregula.

1. In an 11-inch straight-sided skillet, heat the olive oil over medium heat. Add the lamb, sprinkle with salt, and cook until browned all over, 6 to 8 minutes. Add the onion and parsley and continue to cook until the onion is tender, 3 to 5 minutes. Stir in the tomato paste and cook until it begins to brown, about 1 minute. Add 2½ cups of the broth and bring to a boil. Cover, reduce the heat to low, and cook, stirring occasionally, until the lamb is tender, about 1 hour. Add the basil and zucchini and cook until the zucchini is just tender, about 5 minutes.

2. Meanwhile, bring a large pot of well-salted water to a boil. Add the fregula and cook until just al dente, about 1 to 2 minutes less than the package instructions call for. Drain the fregula well and transfer to the pan with the lamb ragù. Add the remaining ½ cup of broth and continue cooking until the fregula is al dente, 1 to 2 minutes. Season to taste with salt and sprinkle with the pecorino cheese.

SERVES 4

1 tablespoon (15 ml) extra-virgin olive oil

1 pound (453 g) lamb meat, cut into ⅛-inch cubes

Kosher salt

1 medium yellow onion, finely chopped

¼ cup finely chopped flat leaf parsley

1 tablespoon tomato paste

3 cups (750 ml) low-salt chicken broth

¼ cup thinly sliced basil

1 small zucchini, cut into quarters and then thinly sliced

12 ounces (340 g) fregula

1 tablespoon freshly grated pecorino cheese

Casu Marzu

The Sardinian shepherding culture has forged a distinct pastoral cuisine, and sheep's-milk cheese has become one of the region's most important and recognized products. While the slightly aged Pecorino Sardo leaves the island's shores for the global market, there exists a coveted and cherished version of pecorino that Sardinians keep for themselves. Casu Marzu begins as wheels of regular pecorino, but instead of ripening in cellars, they are left out in the open with a small incision made in the top to attract cheese flies. These flies lay eggs, which hatch larvae that permeate the interior, living their lives feeding on the cheese and producing enzymes to stimulate fermentation. They have done their job when the inside of the once hardened pecorino becomes creamy, soft, and "alive" with feeding maggots. An acquired taste, the pungent, eye-watering, tongue-singeing flavor is perhaps most appreciated by locals.

As stricter EU regulations have made their way into Italy, Casu Marzu has become a banned product, with government officials declaring the cheese unsafe to consume given the risk of intestinal larval infection in humans. Although now illegal to produce and purchase, the cheese sells on the black market of the island, which has added to its mystique and brought the cheese even closer to the hearts of many Sardinians. Sa Mandra always has a few wheels "aging" for adventurous guests who request a bite of the "living" cheese.

Sardinian Spiced Cookies
PAPASSINI

These popular spice-scented diamond-shaped Sardinian cookies are chock-full of delicious ingredients. They are found all over the island and vary from town to town, and family to family. They are often prepared with a sugary glaze on top, but Rita made hers plain so the subtle nuance of spice would shine through.

1. Position a rack in the center of the oven and heat the oven to 350°F (180°C). Line two baking sheets with parchment paper

2. In a stand mixer with the paddle attachment, combine all of the ingredients and mix on low speed until the dough comes together. Turn the dough out onto a clean work surface and roll out to a rectangle ¼ inch thick. Cut the dough lengthwise into 2-inch strips, then cut each strip on a diagonal into about 1½-inch diamonds. Transfer the diamonds to the baking sheet. Reroll the scraps, cutting out more cookies, until all the dough is used.

3. Bake the cookies, one baking sheet at a time, until golden brown and set, 15 to 20 minutes. Transfer to a wire rack and let cool.

MAKES ABOUT 30 COOKIES

2 cups (250 g) all-purpose flour

6 tablespoons (¾ stick) unsalted butter

5 tablespoons (73 ml) whole milk

1 large egg

¼ cup chopped slivered almonds

¼ cup raisins

1 teaspoon orange zest

½ teaspoon aniseed

½ teaspoon baking powder

¼ teaspoon baking soda

¼ teaspoon cinnamon

Pinch of table salt

Meringue Cookies with Almonds and Orange Zest

MARIPOSSOS

3 large egg whites

1/3 cup granulated sugar

2 tablespoons slivered almonds, toasted and finely chopped

2 teaspoons orange zest, toasted

In these cookies the airy, soft, and chewy texture of the meringue gets a punch of flavor from the orange zest—and a little added crunch from the ground almonds. The consistency is so appealingly addictive, you can't eat just one.

1. Position a rack in the top and bottom third of the oven and heat the oven to 175°F (80°C).

2. In a stand mixer fitted with the whisk attachment, beat the egg whites on medium-low speed until soft peaks form. In a steady stream, gradually pour in the sugar, increase the speed to medium-high, and continue beating to glossy stiff peaks. With a spatula, gently fold in the almonds and orange zest. Spoon the mixture into a large pastry bag fitted with a plain ½-inch tip.

3. Line two baking sheets with parchment paper and pipe out the meringues into rounds 1½ inches in diameter by ½ inch high. Space the meringues ½ inch apart.

4. Bake the meringues until they are dry and crisp and not at all browned, about 2½ to 3 hours. Turn off the oven, leave the door slightly ajar, and let the meringues sit in the oven until cooled, about 1 hour. Remove them from the oven and gently lift the meringues off the parchment. Store in an airtight container for a few days.

MAKES ABOUT 30 COOKIES

Rita making a sweet Sardinian cheese

SA TIRIA

The arid, sun-scorched countryside of southwestern Sardegna feels more like a vast desert than a Mediterranean island. Driving down from the north through the open expanse of brown earth and cacti, you feel hundreds of miles away from the sea. When the sparkling blue waters appear on the horizon, the contrasts of Sardegna's topography become apparent. Here, set into the foothills, lies the agriturismo of Sa Tiria, whose bright yellow-and-orange facade pops against a backdrop of barren landscape. The Ledda family have been farming their land for over one hundred years, raising sheep for cheese, pigs for meat, and bees for honey; most recently they've planted olive trees for oil. In 2001 daughters Linda and Franca spearheaded the decision to expand the farm by opening their doors to tourism. An immense restoration of the entire property ensued, with the farm transforming itself into a holiday destination aptly named Sa Tiria, for the wildflowers that proliferate in the area. Today the Leddas continue their agrarian traditions, with the men tending the farm while the women work in the kitchen and run the agriturismo and restaurant.

The constantly changing menu reflects the seasons and cycles of the farm. Antipasti range from an assortment of the farm's cured hams and dried salamis, to a selection of hard and soft sheep's-milk cheeses, to a more humble sauté of pig's heart and liver. Primi are simple and typical of the island. Each morning Linda, Franca, and their mother, Lucia, decide what to serve and prepare the ingredients. Typically, after mixing semolina flour and water (eggs are obsolete in Sardinian pasta) and passing the dough through a pasta machine, the women convene at a long table to sit, talk, shape, and cut the pasta for that night's meal. Effortlessly, they shape tiny gnocchetti rolled over small ridged wooden boards, make pillows of sheep's ricotta and nutmeg filled ravioli, and roll out long, thin sheets of the golden dough for a vegetable lasagna. Spit-roasted suckling pig, braised lamb, and chicken roasted in a wood oven are a few of the second courses offered, all of which are reared on the farm directly behind the restaurant. Desserts reflect Sa Tiria's agricultural activity and are made with a fermented sheep's cheese. The soft cheese becomes a filling for two of the house specialties: pardulas and sebadas, both made from a flaky, lard-enriched dough. The main difference lies in their shapes, with pardulas resembling stars with a center filling, and sebadas crimped rounds that look like circular calzones. Once fried, they both get a drizzle of the farm's own honey made from the various flowers and herbs that grow wildly around the property. The Ledda family's preference to serve only what they can produce validates Sa Tiria as an authentic taste of rural Sardegna.

Southern coast of Sardegna

Sautéed Peppers with Green Olives
PEPPERONI CON OLIVE VERDI

1 tablespoon (15 ml) extra-virgin
    olive oil

4 long green peppers, seeded and
    cut into thin strips

2 medium yellow bell peppers,
    seeded and cut into thin strips

1 large yellow onion, thinly sliced

¾ cup pitted green olives

Kosher salt

¼ cup thinly sliced basil leaves

½ tablespoon (7 ml) white wine
    vinegar

At Sa Tiria they serve this dish as an appetizer with wedges of aged pecorino cheese. The addition of white wine vinegar at the end of cooking gives a touch of brightness to the dish.

In a large straight-sided skillet, heat the olive oil over medium heat. Add the peppers, onion, olives, and a generous pinch of salt. Cook, stirring occasionally, until the vegetables are tender but not breaking down, 10 to 12 minutes. Stir in the basil and the vinegar and cook until the flavors meld together, about 1 minute. Serve warm.

SERVES 4–6

PANE CARASAU—CARTA DI MUSICA

The regional differences in gastronomic products in Italy are varied and vast, and their origins speak of time and place. Bread is as important a staple as any in the Italian repertoire, and there are many different types across the country. One of the most unusual hails from Sardegna and is actually more akin to a cracker. Pane carasau are thin, unleavened sheets of dough, which are baked twice to achieve their crispy texture. Their origins date back to Sardegna's shepherding heritage, when wives would pack stacks of the bread for their husbands to take while they were away from home tending to their animals. Easy to transport and with a long shelf life, the flat disks were also used as a type of edible plate. At Sa Tiria baskets of pane carasau are set on tables at dinner and refilled throughout the evening, as the bread's pleasing texture proves to be quite addictive. Pane carasau can also be moistened and softened to resemble sheets of pasta, which are layered with sauce and baked.

Braised Lamb Shanks with Sundried Tomatoes

STINCO DI PECORA BRASATO

Sardegna's identity lies within its shepherding culture. Sheep have always been raised here for cheese—and of course for their meat, whose slightly gamy flavor plays an integral role in the island's cuisine. At Sa Tiria they dry their own tomatoes every summer in the scorching sun. The tomatoes are then preserved in olive oil and used throughout the year to add a summer touch to dishes. In this recipe the tomatoes are added to a long-simmered braise with lamb shanks, making for a fork-tender taste of Sardegna.

1. Position a rack in the bottom third of the oven and heat the oven to 300°F (160°C).

2. Season the lamb shanks generously with salt. In an 8-quart heavy-duty pan or Dutch oven, heat 2 tablespoons of the olive oil over medium-high heat. Add the lamb shanks, in batches if necessary, and brown all over, 3 to 5 minutes per side. Remove the shanks from the pan and set aside on a plate. Add the remaining 1 tablespoon of oil with the onion, garlic, parsley, sun-dried tomatoes, and a pinch of salt; cook until the onions are tender, 5 to 7 minutes. Add the wine to the pan and reduce by half, about 2 minutes.

3. Return the lamb shanks to the pan along with the tomatoes, bay leaves, and 1 cup of water. Bring the mixture to a boil. Cover the pan and place in the oven. Cook until the meat is fork-tender, about 2½ to 3 hours. Discard the bay leaves. Put the shanks on a platter and spoon over the sauce. Serve immediately.

SERVES 4

4 lamb shanks (12–16 oz./340–453 g each)

Kosher salt

3 tablespoons (44 ml) extra-virgin olive oil

1 medium yellow onion, cut into small dice

1 clove minced garlic

2 tablespoons fresh flat-leaf parsley

12 sun-dried tomatoes (not packed in oil)

3/4 cup (175 ml) red wine

1 28-ounce (793 g) can whole plum tomatoes, crushed with their juices

2 bay leaves, preferably fresh

Semolina Gnocchi

GNOCCHETTI SARDI

1 recipe semolina pasta dough (see "Basic Recipes")

2 tablespoons (30 ml) extra-virgin olive oil

1 medium yellow onion, minced

1 carrot, minced

½ cup minced flat-leaf parsley

Kosher salt

1 28-ounce (793 g) can whole plum tomatoes, crushed with their juices

2 bay leaves

Pecorino cheese, for serving

These small semolina gnocchi are a traditional dish of Sardegna and often are served at weddings or other celebrations. To get their ridges they are either cut by hand and rolled across a small wooden board with deep grooves, or rolled into strips and passed through a hand-cranked pasta maker, which speeds up the process of making these tiny dumplings. In this case the back of a cheese grater will suffice to ensure nice ridged pasta.

1. Cut the dough into eight pieces and roll out each piece into a ¼-inch-thick rope. Using a knife or pastry cutter, cut each rope into ¼-inch pieces. With your thumb, roll each piece of cut dough against the back of a handheld cheese grater. Transfer the gnocchi to a baking sheet lined with parchment paper.

2. Heat the oil in a 5- to 6-quart pot over medium-low heat. Add the onion, carrot, parsley, a pinch of salt, and 2 tablespoons of water. Cover the pot with a lid and cook the mixture, stirring occasionally, until tender but not at all browned, 10 to 12 minutes. Adjust the heat as necessary to ensure that the mixture does not brown. Add the tomatoes, and bring to a vigorous simmer. Reduce the heat to maintain a gentle simmer and add the bay leaves. Cook uncovered until the sauce is thick and flavorful, 45 to 50 minutes. Season to taste with salt.

3. Bring a large pot of well-salted water to a boil. Drop the gnocchi into the boiling water and cook until they are al dente, 12 to 15 minutes. Drain the pasta well and toss with the sauce. Serve with freshly grated pecorino cheese.

SERVES 6–8

Semolina gnocchi

Fried Cheese Ravioli

SEBADAS

Sebadas, renowned in Sardegna, are a divine dessert. Sweetened sheep's-milk cheese is stuffed into a ravioli that gets fried and then served warm drizzled with honey. What better way to end a meal?

For the dough:

4 cups (500 g) semolina flour

1 cup lard or shortening

Pinch of Kosher salt

1¼ cups (300 ml) water

For the filling:

2 cups (500 g) ricotta

½ cup finely grated pecorino

1 cup (125 g) all-purpose flour

¼ cup granulated sugar

1 egg yolk

1 tablespoon lemon zest

1 tablespoon honey, preferably
   wildflower, plus more for serving

1 tablespoon (15 ml) Sambuca
   (optional)

To assemble:

Canola oil for frying

Granulated sugar or honey
   for serving

1. **Make the dough:** In a stand mixer fitted with a dough hook, add the flour, lard (or shortening), and salt, and mix on low to combine. Gradually add the water and then increase the speed to medium. Mix until a dough forms around the hook, 5 to 8 minutes. Transfer the dough to a clean work surface and cover with a kitchen towel.

2. **Make the filling:** Add the cheeses to the stand mixer fitted with the dough hook and mix on low to combine. Add the flour, sugar, egg yolk, lemon zest, honey, and Sambuca, if using, and mix on medium speed until combined.

3. **To assemble,** divide the dough into four pieces. Using a pasta machine, roll out the dough, starting with the widest setting and then going down until the dough is ⅛ inch thick. Cut out disks from the dough using a 4-inch-round cookie cutter. Place a heaping tablespoon of filling in the center of a round, and then top with another disk of dough. Press out any air between the filling and the dough and then crimp the edges. Finish with the remaining dough and filling. Transfer the ravioli to a baking sheet lined with parchment paper; cover with parchment paper and refrigerate until ready to cook.

4. Heat a 12-inch cast-iron skillet with ½ inch of canola oil until shimmering hot. Add the ravioli and fry until golden brown, 2 to 3 minutes. Flip the ravioli over with a slotted spatula and continuing frying until nicely golden, about 2 minutes. Transfer to a paper-towel-lined plate and continue fry-

ing the remaining ravioli. Serve on individual plates, drizzling each with ½ tablespoon honey. Serve immediately.

SERVES 10

Forming the Sebadas

IL MUTO DI GALLURA

Deep in the rugged inland terrain of northeastern Sardegna lived Il Muto di Gallura—a legendary and elusive bandit. Born a mute in the early 1800s, the notorious outlaw wandered throughout the Galluran landscape of dense cork tree forests dispersed with enormous outcroppings of granite and jagged mountainous peaks. His Robin Hood–like tale of brutal crimes against the rich to help the poor resonates with romanticism and pride. Husband and wife Gianfranco and Francesca Serra named their agriturismo in honor of the voiceless hero. Gianfranco, a rugged, Clark Gable type, boasts that he is a descendant of the Muto, and through the work on his farm, he carries on his lineage in the same hills and fields where the legend lived. Seeing Gianfranco mounted on horseback with a rifle slung across his back as he canters through the desolate 300 acres that surround his farm feels like stepping back into another era, and to a time of fabled stories and elusive bandits.

The hills that surround Muto di Gallura teem with partridge, hare, quail, and wild boar, and Gianfranco has turned the property into a private hunting reserve for guests of the agriturismo. Avid hunters from all over the world come to bag native species and to taste the flavors of the countryside. In an all-terrain-outfitted Land Rover Defender Ninety, Gianfranco engages the four-by-four and drives his guests down twisting, overgrown, pockmarked dirt roads, deep into the woods for the hunt. Any animals taken that day are brought back to the kitchen, where Francesca prepares classic renditions of wild game.

The Pernice Sardo, an indigenous partridge found only in the hills encircling Muto di Gallura, is a prized catch. Francesca prepares the bird either braised whole with olives and herbs or slowly simmered in a tomato sauce and then shredded and served over pasta. Wild boar gets simmered in red wine and herbs to shed some of its gaminess, and then cooked in a sweet-and-sour sauce that tantalizes the senses with its contrasting flavors.

Francesca's signature first course—Zuppa Gallurese, a bread soup made from veal stock and layered with cheese and herbs—reflects the province's impoverished past. The no-frills appearance of the soggy bread bowls presented before guests does little justice to the sublime deliciousness of the dish, where each mouthful contains a bite laden with the rich long-cooked broth and melting sheep cheese. While many of Francesca's recipes speak of a peasant diet from another era, the abundance and generosity with which they are served speaks of the prosperity of present-day Sardegna. Surrounded by vineyards, the farm's award-winning wine accompanies meals, and its easy-drinking quaffable qualities work their mellow charm on diners throughout the evening. A shot of homemade mirto—liquor distilled from tiny purple berries of myrtle bushes—always finishes the night. Its smooth, subtle kick helps aid in digestion. In tune with the Serra's hospitable spirit, entire bottles are left on the table for guests to sip on long into the night.

Sardinian landscape

Gallurese Soup
ZUPPA GALLURESE

For the broth:

2 pounds (900 g) beef marrow
  bones

1 pound (453 g) scrap veal meat or
  shanks

1 large yellow onion, quartered

2 carrots, cut into 2-inch pieces

2 stalks celery, cut into 2-inch pieces

1 tomato, cut in half

3 sprigs parsley

¼ cup lightly packed basil leaves

2 bay leaves, preferably fresh

For the soup:

½ cup grated pecorino

½ cup grated Parmigiano Reggiano

¼ cup chopped flat-leaf parsley

Freshly ground black pepper

1 loaf rustic Italian bread, cut into
  ¼-inch-thick slices and left out for
  2 days to dry

1 cup grated scamorza or provolone
  cheese

No dish represents the region that Il Muto di Gallura calls home more than Zuppa Gallurese. This rustic, bread-based soup illustrates the once bucolic lifestyle of the area, and is deeply rooted in its food culture. Nourishing both the appetite and the soul, this hearty fare will be best appreciated during the cold winter months. The meat broth is essential to this recipe, so making the rich stock is definitely recommended over store-bought broth.

1. **Make the broth:** In a large (8-quart) stockpot, add the marrow bones, veal scraps (or shanks), onion, carrots, celery, tomato, parsley, basil, and bay leaves. Cover with water by 1 inch. Bring the mixture to a boil over medium heat. Reduce the heat to just a simmer and skim any scum from the surface. Cook until the broth is rich and flavorful, 4 to 5 hours. Strain the broth through a fine-mesh sieve into a clean pot. Let cool and then refrigerate overnight. Skim away any fat that has consolidated on the surface.

2. Position a rack in the center of the oven and heat the oven to 350°F (180°C).

3. **In a small bowl,** mix together the pecorino, Parmigiano, parsley, and a few generous grinds of black pepper. In an 8 x 8-inch baking dish, arrange a layer of bread so that it covers the bottom of the dish. Sprinkle with half the scamorza cheese, then a quarter of the pecorino mixture. Top with another layer of bread, then the remaining scamorza cheese, and a quarter of the pecorino mixture. Top with a final layer of bread and then with the remaining half of the pecorino mixture. Pour in enough broth just to come up the side of the final layer of bread, about 4 cups. Bake in the oven until the top is a deep brown and crusty and the liquid has been absorbed, about 1 hour. Transfer to a rack and let cool slightly before serving.

SERVES 4–6

Hunter-Style Quail
QUAGLIE ALLA CACCIATORA

The game reserve at Muto di Gallura teems with game birds, and a prized species is an indigenous partridge (pernice sarda) known to live only on the island. During our stay there, a German couple spent an afternoon hunting and bagged a number of these small gray-and-red birds, and that evening we were able to taste this Sardinian delicacy. We have adapted the recipe to use quail, and this can also be served over pasta or a bed of rosemary-roasted potatoes.

1. Season the quail generously with salt and pepper. Heat 2 tablespoons of the oil in an 11-inch straight-sided pan over medium-high heat. Add half of the quail and sear until golden brown all over, about 8 minutes total. Transfer to a plate and repeat with the remaining quail.

2. Add the sausage, onion, garlic, and a pinch of salt to the pan; cook until browned and the onions are tender, 5 to 6 minutes. Add the bay leaves, parsley, and rosemary to the pan and cook until fragrant, about 1 minute. Add the white wine, and scrape up any browned bits from the bottom of the pan with a wooden spoon. Bring the wine to a boil and reduce by half. Add the quail and the broth, reduce the heat, and cook, covered, at a gentle simmer until the quail is cooked through and almost falling off the bone, about 20 minutes. Add the olives and cook until the flavors have melded together, about 2 minutes. Season to taste with salt and pepper. Transfer to a platter and serve.

SERVES 4

8 quail (about 5 oz./141 g each)
Kosher salt and freshly ground black pepper
¼ cup (59 ml) extra-virgin olive oil
3 ounces (85 g) sweet Italian sausage, casing removed, crumbled
1 medium yellow onion, thinly sliced
2 cloves garlic, smashed
4 fresh bay leaves
1 tablespoon chopped flat-leaf parsley
2 teaspoons rosemary
½ cup (118 ml) white wine
½ cup (118 ml) low-salt chicken broth
¼ cup chopped green olives

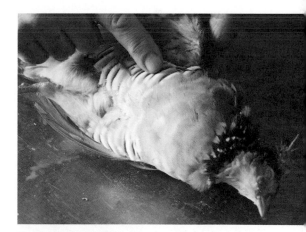

Sardinian partridge

Sweet-and-Sour Wild Boar

CINGHIALE IN AGRODOLCE

1½ pounds (680 g) wild boar roast,
cut into 2-inch pieces

1 bottle (750 ml) dry red wine

2 tablespoons (30 ml) extra-virgin
olive oil

1 small yellow onion, cut into fine dice

1 carrot, cut into fine dice

1 stalk celery, cut into fine dice

Kosher salt

2 fresh bay leaves

1 clove garlic, smashed

1 28-ounce (793 g) can whole plum
tomatoes, chopped with their
juices

1 tablespoon granulated sugar

1½ (22 ml) tablespoons white wine
vinegar

The native wild boar that roam throughout the thick brush surrounding Muto di Gallura are a smaller and leaner species than those on the mainland, and live on a diet of foraged acorns. When hunters visiting the farm take one of the wild pigs, this is a classic preparation that tames some of the gamy flavors of the meat. A good source for wild boar in the States can be found at www.dartagnan.com.

1. Put the boar and red wine in a 4-quart pot and bring to a boil over medium-high heat. Reduce the heat to maintain a gentle simmer and cook to remove some of the gaminess of the meat, about 15 to 20 minutes. Drain the meat and discard the wine.

2. In an 11-inch straight-sided pan, heat the oil over medium heat. Add the onion, carrot, celery, and a pinch of salt; cook until the vegetables are tender and just beginning to brown, about 6 minutes. Add the bay leaves and garlic and cook until fragrant, 2 to 3 minutes. Add the boar plus a generous pinch of salt and cook for 2 minutes. Add the tomatoes to the pan and bring the mixture to a boil. Reduce the heat, cover partially, and cook at a very slow simmer until the meat is very tender, 2 hours. If the sauce becomes too thick, add a bit of water to it.

3. Stir in the sugar and the vinegar and season to taste with salt. Serve on a large platter.

SERVES 4

Wild boar with baby

Individual Fig Cakes
TORTINE DI FICHI

1½ cups granulated sugar, plus 2
  tablespoons

3 large eggs

2½ cups (300 g) all-purpose flour

1 cup (240 ml) whole milk

1 cup (240 ml) vegetable oil

¾ teaspoon baking powder

½ teaspoon baking soda

½ teaspoon (2 ml) vanilla extract

4 ounces (115 g) fresh figs, cut into
  ½-inch pieces

When Francesca makes these simple cakes, the rest of the staff seem to drop by and sneak a few from her tray. These quick little cakes can be made with a host of different flavorings. Mix in cut-up apple pieces or toasted chopped almonds, or mix half of the batter with cocoa powder and then fill the cups with 1 tablespoon of each batter.

1. Position a rack in the center of the oven and heat the oven to 350°F (180°C).

2. In a stand mixer fitted with the whisk attachment, add 1½ cups of the sugar and the eggs. Beat on medium speed until thick and creamy, 3 to 4 minutes. Add the flour, milk, and oil and continue to beat until combined. Add the baking powder, soda, and vanilla and mix until the batter is smooth and thick.

3. Meanwhile, in a small bowl toss the figs with ½ tablespoon of sugar and let sit for a few minutes to extract the juices.

4. With a rubber spatula, fold the figs into the batter. Portion out the batter into standard-size individual muffin liners (no pan needed) arranged on a baking sheet. Sprinkle the remaining 1½ tablespoons sugar over the tops of the cakes.

5. Bake in the oven until the tops are golden brown and a toothpick inserted into the center comes out clean, 20 to 25 minutes.

MAKES 16 CAKES

Just-plucked ripe figs

CHAPTER 2

TOSCANA (TUSCANY)

GUALDO DEL RE

In the early 1980s Nico Rossi decided to take a risk and pursue his passion for wine by planting vines around his family's farm. He was among a small group of pioneers who saw great potential for the production of quality wine in the Maremma area of Tuscany. In addition to the typical varietals of Sangiovese and Vermentino, Nico believed that other grapes could thrive in the temperate climate and rich soil, made up of clay, sand, and stone. In addition to the classic Bordeaux varietals of Merlot and Cabernet Sauvignon, he planted an array of vines including Pinot Blanc, Gewürztraminer, and Pinot Noir. His newfound vineyard was named Gualdo del Re in homage to a German landlord who once owned the property for hunting.

As the Rossi family was beginning to establish a name for their vineyard, the region of Tuscany was becoming a jewel in the eye of visiting tourists. Soon the Maremma was following a similar path in the world of wine lovers, gaining international recognition for soft structured big reds. The Super Tuscan was born, and the Maremma became recognized around the globe as home to some of the most sought-after and expensive wines coming out of Italy. Gualdo del Re began reaping the benefits of its winemaking gamble. In a similar fashion to the birth of his winery, Nico opened an agriturismo, believing in the beauty of the Bolgheri section of the Maremma as one of Tuscany's most enchanting landscapes, catering to oenophiles looking to vacation among some of the most coveted vines in all of Europe. With a backdrop of soft rolling vine-covered hills dotted with cypress-lined streets and quaint stone villages, the farm offers individual apartments as well as one of the area's most sophisticated restaurants.

With a seasonal menu selected for its ability to pair with the estate's wines, the winery offers diners an unparalleled way to experience all ten of the vineyard's selections. From simple and fresh whites to sleek and modern reds, the gauntlet of Nico's nectar is vast and impressive. All emerge from a minimalist cellar made up primarily of large oak casks where nearly all the wine rests before bottling. A row of French barrique barrels are designated for the more refined reds, like a 100 percent Sangiovese, the winery's signature wine, aptly named Gualdo del Re. Gian Maria Margelli has been working as the in-house chef since the doors opened in 2005. With work experiences in

TOSCANA (TUSCANY)

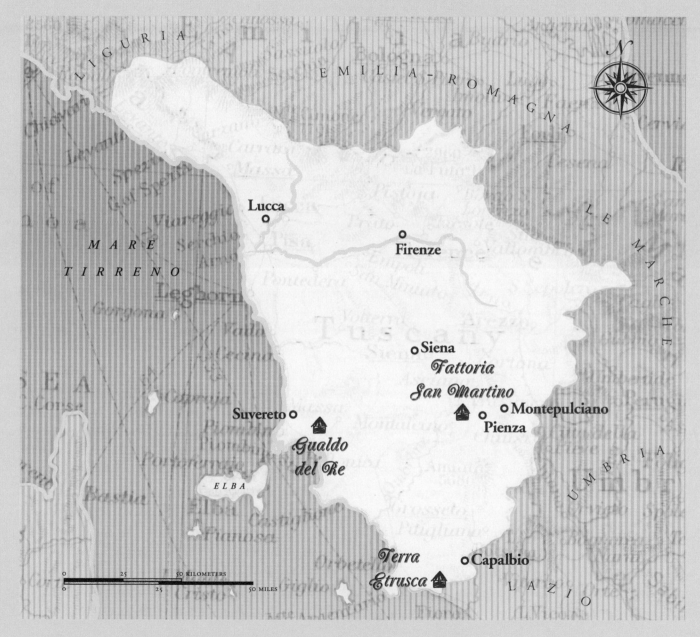

LIGURIA

EMILIA-ROMAGNA

LE MARCHE

Lucca

Firenze

MARE
TIRRENO

Tuscany

Siena

*Fattoria
San Martino*

Suvereto

Montepulciano

Pienza

*Gualdo
del Re*

ELBA

UMBRIA

*Terra
Etrusca*

Capalbio

LAZIO

0   25   50 KILOMETERS
0   25   50 MILES

both Italy and Paris, he has developed a cooking style that respects Tuscan roots with influences from abroad. Refined and elegantly plated, his meals complement the slightly extracted and concentrated flavors of Nico's wines, which in turn enhance Gian Maria's cuisine. The a la carte menu includes both a rustic traditional section as well as one with elevated, gourmet variations, each always highlighting local products. Briny mussels from the nearby Tyrrhenian Sea stewed with chickpeas, and baby gnocchi with candied tomatoes and smothered cabbage pair with Eliseo, a Pinot Bianco and Vermentino blend, or the Eliseo Rosato, a rosé of Sangiovese and Merlot. Native white truffles and wild game animals, such as boar and hare, which populate the dense forests of the Maremma, play heavily into the cuisine. A delicate lightly poached chicken breast takes on earthy tones with a truffle foam, pappardelle are sauced with a hearty wild boar ragù, and

a rabbit terrine is infused with rosemary; all are matched with bolder reds, like Rennero, a barrique-aged Merlot, or a Pinot Noir named Senzansia.

From his tiny two-man kitchen, Gian Maria prepares everything from scratch, and desserts are one of his specialties. The bitterness of dark amber caramel cuts the cloyingness of white chocolate in a molten cake; a two-layered mousse is sweetly caffeinated with espresso and vanilla. All are paired with Amansio, a red sweet dessert wine made from dried grapes. With Gualdo del Re's earned reputation as one of the Maremma's top producers, Gian Maria's cooking is beginning to achieve separate accolades, which are bringing diners out to visit the vineyard for dinner. This has brought even greater exposure to Nico's wines and has brought the winery/agriturismo into the forefront as an ambassador for the area and as a destination to experience an elevated level of Tuscan food and wine.

Rosemary-Infused Rabbit Rillettes
TERRINA DI LEPRE CON ROSMARINO

Wild hare are quick and elusive and a prized catch of Tuscan hunters. Dogs are sent into the thick brush of the rabbits' habitat, chasing them into an open clearing where the hunters will have a better chance for a good shot. The wild hare of the Tuscan forests are slightly gamy and strong in flavor; we have adapted this recipe for domestically raised rabbits. Be sure to serve this with plenty of grilled or toasted crusty bread.

1. Cut the rabbit into six pieces. Discard the saddles, livers, and kidneys.

2. Put the onion, carrot, celery, thyme, parsley, and bay leaf into an 11-inch straight-sided sauté pan. Add the white wine and 4 cups of water. Bring to a boil over medium-high heat.

3. Add the rabbit to the pan, return to a boil, and skim any scum that rises to the surface. Season with a pinch of salt and a few grinds of pepper and simmer, covered, over low heat until the rabbit is falling off the bone, about 2 hours.

4. Remove the rabbit pieces and set aside to cool. Once they're cool enough to handle, debone them and finely shred the meat. Transfer the meat to a large bowl and add the rosemary.

5. In a small saucepan, melt the duck fat without heating it too much. Add the duck fat to the rabbit little by little, working in each addition well with a spatula to bind the meat together. Season to taste with salt and pepper and serve immediately.

SERVES 6

1 rabbit, (about 2 pounds/1 kg)

1 yellow onion, chopped

1 carrot, peeled and chopped

1 stalk celery, chopped

2 sprigs thyme

2 sprigs flat-leaf parsley

1 bay leaf

5 tablespoons (73 ml) white wine

Kosher salt and freshly ground
   pepper

1 sprig rosemary, minced

2 ounces (60 g) duck fat

Rabbit rillette

Gnocchetti with Savoy Cabbage and Candied Cherry Tomatoes

GNOCCHETTI CON CAVOLO CAPPUCCIO
E POMODORINI CANDITI

2 pounds (900 g) russet
potatoes

Kosher salt

2⅓ cups (310 g) all-purpose flour

1 large egg

2 14-ounce cans (400 g)
cherry tomatoes, strained

¼ cup (59 ml) extra-virgin olive oil

1 teaspoon granulated sugar

2 whole cloves

2 sprigs thyme

1 3-inch cinnamon stick

1 1-inch strip orange zest

1 1-inch strip lemon zest

1 sprig rosemary

¼ teaspoon whole black
peppercorns

1 small head savoy cabbage,
shredded

In this dish, slow-roasting canned tomatoes with sugar concentrates their natural sweetness, while the spices add depth and intensity. The candied sauce pairs nicely with the long-simmered cabbage, making for a hearty cold-weather meal that is filling and nourishing when served over tiny gnocchi, known as gnocchetti.

1. Put the potatoes in a large pot and cover with cold water. Bring to a boil over medium heat, salt liberally, and cook at a gentle simmer until the potatoes are extremely tender, 35 to 40 minutes. Drain the potatoes and peel them while they're still warm. Pass through a ricer onto a clean work surface.

2. Sprinkle the flour evenly over the potatoes and form into a well. Crack the egg into the well and gently beat with a fork. Gradually incorporate the egg into the flour. Knead the dough until smooth.

3. Cut the dough into four even pieces. Roll each piece out into a ¼-inch-thick rope, and then cut each rope with a pastry cutter or knife into ¼-inch pieces. Place the gnocchetti on sheet pans lightly dusted with flour, cover, and refrigerate until ready to use.

4. Position a rack in the center of the oven and heat the oven to 325°F (170°C).

5. In a large bowl, toss the tomatoes with 2 tablespoons of the oil, 1 teaspoon salt, and the sugar; toss well to combine. Arrange the tomatoes in a single layer in an 8 x 8-inch baking dish. Make a bundle out of cheesecloth with the cloves, thyme, cinnamon stick, orange zest, lemon zest, rosemary, and peppercorns. Add this bundle to the pan and roast in the oven until the tomatoes are tender and beginning to caramelize, about 1½ hours. Remove from the oven and discard the bundle.

6. In a 12-inch skillet heat the remaining 2 tablespoons oil over medium heat. Add the cabbage and a generous pinch of salt and cook, stirring often, until the cabbage wilts, about 5 minutes. Add the tomatoes and ¼ cup of water, cover, and reduce the heat. Cook, stirring often, until the cabbage is meltingly tender, 35 to 40 minutes. Season to taste with salt.

7. Bring a large pot of well-salted water to a boil. Drop the gnocchetti into the water and cook until they begin to rise to the surface, about 3 minutes. Transfer the gnocchetti to the cabbage with a slotted spoon or spider, and toss to combine.

8. Serve the gnocchetti in individual shallow bowls.

Aging bottles of wine

Curly Tagliatelle with Braised Boar and Olives

RICCIOLINE AL GERME DI GRANO STUFATO DI CINGHIALE E OLIVE

2 tablespoons (30 ml) extra-virgin olive oil

1 yellow onion, finely chopped

1 stalk celery, finely chopped

Kosher salt and freshly ground black pepper

1½ pounds (680 g) wild boar shoulder roast, boned, trimmed of fat, and cut into ¼-inch pieces

½ cup (118 ml) dry red wine

2 cups (473 ml) homemade or low-sodium beef broth

2 fresh bay leaves

2 sprigs flat-leaf parsley

1 sprig rosemary

½ cup small green olives, pitted and halved

½ cup small black olives, pitted and halved

1 pound (453 g) mafaldine pasta (or tagliatelle)

Tuscany lays claim to some of Italy's most respected hunters. Provinces throughout the region have hunting clubs that meet during the fall months to seek out the prized wild boar that live deep in the woods. Many clubs have their own lodges, and after mornings spent hunting cinghiale the men reconvene for a communal lunch where wine and grappa flow as stories about the hunt are relived. A good source for wild boar in the states can be found at www.dartagnan.com.

1. Heat the oil in a 12-inch straight-sided skillet over medium heat. Add the onion, celery, and a generous pinch of salt; cook until the vegetables are tender and beginning to brown, 6 to 8 minutes. Season the wild boar meat all over with ½ teaspoon salt and ¼ teaspoon pepper. Add the wild boar to the pan and cook, stirring occasionally, until nicely browned all over, 5 to 7 minutes. Pour in the red wine, bring to a simmer, and then reduce by half. Pour in the broth, bring to a boil, and reduce the heat to maintain a gentle simmer. Add the bay leaves, parsley, and rosemary, cover the pan, and cook until the meat is fork-tender, 45 to 50 minutes. Stir the olives into the sauce and cook until warmed through. Season to taste with salt and pepper.

2. Meanwhile, bring a large pot of well-salted water to a boil over medium-high heat. Drop the pasta and cook until al dente, following the package instructions. Reserve ¼ cup of water and then drain the pasta.

3. Toss the pasta together with the sauce, adding any reserved pasta water as needed if the sauce seems dry.

SERVES 6

Guinea Fowl with Grape Sauce
FARAONA ALL'UVA

The fall grape harvest brings the fruit into the kitchen, their juices adding a hint of autumn sweetness to dishes. Often overlooked in savory preparations, grapes add a sugary kick we really like.

1. Put 12 ounces of the grapes into a food processor or blender and puree until juicy and smooth, about 30 seconds. Strain the grapes through a fine-mesh sieve, pressing down on the solids to extract the juice; you will need ½ cup of juice.

2. Remove 2 tablespoons of fronds from the fennel and set aside. Cut the fennel in half lengthwise and remove the core. Thinly slice the fennel; toss with 1 tablespoon of the oil and season with salt and pepper. Set aside.

3. Season the guinea fowl or chicken thighs all over with salt and pepper. Heat the remaining 2 tablespoons of oil in a large straight-sided skillet over medium-high heat. Add the guinea fowl and cook to a deep golden color, turning to brown all over, about 8 minutes. Add the garlic and pepper flakes and cook until fragrant, about 1 minute. Pour in the grape juice and bring to a boil, then reduce the temperature and cook until reduced by half, about 4 minutes. Add ½ cup of water, the sage leaves, carrot, peppercorns, and reserved fennel fronds; cover and simmer until the fowl is almost tender, about 20 minutes. Add the reserved grapes, cover, and continue cooking until the fowl is fork-tender and the grapes begin to break down, about 10 minutes. Remove the lid from the pan and transfer the fowl pieces or chicken legs to a plate. Finish cooking the sauce until it has reduced and thickened, 2 to 3 minutes.

4. Transfer the fennel to a serving platter, top with the guinea fowl, drizzle with the sauce, and serve.

SERVES 4

1 pound (453 g) green seedless grapes

1 small fennel bulb with fronds attached

3 tablespoons (44 ml) extra-virgin olive oil

Kosher salt and freshly ground black pepper

1 guinea fowl (about 3½ pounds/1½ kg), cut into 8 pieces, or 4 whole skin-on chicken legs

2 cloves garlic, unpeeled

Pinch crushed red pepper flakes

8 fresh small sage leaves

1 small carrot, peeled and cut into ¼-inch pieces

½ teaspoon white peppercorns, crushed

Vanilla and Espresso Layered Mousse
MOUSSE ALLA VANIGLIA E CAFFÈ

1½ cups (360 ml) heavy cream

2 teaspoons gelatin, divided

½ vanilla bean

1 cup granulated sugar

6 large egg yolks, divided

¼ cup (59 ml) brewed espresso (or other strong coffee)

Chocolate-covered espresso beans, for garnish

The appeal of this dessert lies in the beautiful presentation as well as in the decadent richness of the creamy mousse. The two layers, one infused with vanilla, the other scented with espresso, offer a pleasant contrast in both flavor and color.

1. In a stand mixer fitted with a whisk attachment, beat the heavy cream on medium-high speed until stiff peaks form, 3 to 5 minutes. Transfer to a bowl and refrigerate.

2. In a small bowl, sprinkle 1 teaspoon of the gelatin over 2 tablespoons of cold water; let soften for 5 minutes.

3. Split the vanilla bean in half lengthwise with a sharp paring knife and scrape out the seeds with the back of the knife. Put the seeds in a small saucepan, along with ¼ cup of the sugar and ¼ cup water. Bring to a simmer over medium heat, stirring occasionally until the sugar dissolves. Stir in the softened gelatin until dissolved.

4. In a medium bowl, beat three of the egg yolks and ¼ cup of the sugar with an electric mixer on medium-high until thick and pale yellow, about 5 minutes. In a steady stream, gradually beat in the gelatin mixture, and continue beating until cooled, about 2 minutes.

5. Fold half of the whipped cream into the egg yolk mixture (put the other half back in the fridge). Divide the mixture among six 4-ounce clear glasses, filling each halfway. Put the glasses on a sheet tray and refrigerate until set, about 2 hours.

6. In a small bowl, sprinkle the remaining 1 teaspoon of gelatin over 2 table-spoons water; let soften for 5 minutes. In a small saucepan, heat the espresso with ¼ cup of the sugar over medium heat until the sugar dissolves. Stir in the gelatin until dissolved.

7. In a medium bowl, beat the remaining 3 egg yolks and remaining ¼ cup of sugar with an electric mixer on medium-high, until thick and pale yellow, about 5 minutes. In a steady stream, gradually beat in the espresso mixture; continue beating until cooled, about 2 minutes. Fold the remaining whipped cream into the egg yolk mixture.

8. Divide the espresso mousse evenly among the glasses, filling them to the top. Cover loosely with plastic and refrigerate for at least 4 hours, or up to overnight. Garnish each mousse with chocolate-covered espresso beans and serve.

SERVES 6

TERRA ETRUSCA

The low-lying pastures of the Maremma are home to living cowboys who herd their cattle in vast expanses of fields that extend to the Tyrrhenian Sea. Once a vast marshland laden with mosquito infestations that spread malaria, today the area is ideal for farming with its nutrient-rich soil. For over ten generations the Veronesi family has forged a living cultivating this land. Their name is attached to a great lineage of farmers that toiled through the disease-ravaged years. Carlo Veronesi, the farm's present owner, is a true cowboy at heart. Never without a wide-brimmed hat and well-worn leather boots, he streamlined farm production to focus on his true passion—raising the sturdy workhorses and long-horned cattle of the area, known as the Maremmana. Underneath his shirt a giant scar lends testament to the dangers of rearing such large beasts. His sons Paolo and Alberto continued their family's agricultural past, establishing the farm as completely organic in 1994 and adding more vegetables to its repertoire. Paolo, very much a non-cowboy type, respects the foundation his ancestors laid, but felt that a change was necessary to bring the family business into the next generation. With the recent opening of the Terra Etrusca agriturismo, the Veronesis are poised to embrace the future. In a minimalist style, they have modernized the family farm, creating a hip setting that includes a market, restaurant, and nine guest rooms, run by a staff of young, energetic Italians who keep the place bustling and alive. Two brightly painted white-and-blue rooms, each pulsing with rock music, house the farm shop and restaurant that look into each other through large glass doors. Wooden shelves are lined with Terra Etrusca's own jarred tomatoes, jellies, soups, and bags of pasta, while a gray-and-white-tiled floor and hanging black metal lamps decorate the market, lending a bright and modern feeling to the space. A large blackboard lists the seasonal all-organic produce, stacked in green plastic bins that spill over with just-picked vegetables, fruit, and herbs. Surrounded by manicured rows of vegetables, vineyards, and fruit trees, it is the idyllic place for shopping for dinner.

Paolo and Alberto have instilled in everyone that works at Terra Etrusca a passion for their farm. They all share a common philosophy and believe in the importance of organic farming and cooking; the lighthearted banter and jokes that pepper the workday come from a shared congenial effort toward a common goal. This has created a close camaraderie among employees, and their love for what they do is evident. All play a role in the rhythm of the farm and help with harvesting, cleaning, canning, and cooking farm products. A small, evolving menu is dictated by the seasons and reflects dishes of the Maremma, based on the vast selection of fresh organic vegetables out the kitchen door. Tuscan kale wilted in a soup with a soft runny egg has peasant origins, fiery just-picked cardoons sautéed with garlic and chile pepper taste of earthy goodness, and pureed broccoli and ricotta filled tortelli, demonstrate the kitchen's proficiency at hand-shaped pasta. Large cuts of steak from Terra Etrusca's own free-range Maremmanan cattle, known for their deep red color and richness, are quickly grilled and dressed with sea salt and olive oil and served with peppery arugula. Rivaling the region's infamous fiorentina T-bone steak from Chianina beef, Terra Etrusca's steaks are deeply marbled and succulent, and diners are presented with raw sides cut to order to showcase the veins of snow-white fat against the crimson meat. Carlo is also a member of a local hunting club that owns land and a lodge deep in the forested countryside, alive with wild boar, hare, and deer. Game animals taken on successful days are brought back to the kitchen and highlighted as specials, often in various sauces for pasta or marinated and braised in red wine and herbs. Desserts are bountiful and include a sponge cake log rolled with apricot jelly, a ricotta mousse drizzled with the farm's own honey, and hazelnut biscotti made with grappa and anise that get dunked in the sweet Tuscan dessert wine Vin Santo. As the Maremma develops into a popular tourist destination for those seeking a more off-the-beaten-path vacation, Terra Etrusca is on the brink of becoming recognized as a premier spot in the area for its chic design and gratifying cuisine. It's certainly a far cry from a once mosquito-laden swampland.

Fresh herbs and shallots

"Cooked Water" with Tuscan Kale and Poached Egg

ACQUACOTTA

2 tablespoons (30 ml) extra-virgin olive oil, plus more for serving

3 small yellow onions, halved and thinly sliced

3 stalks celery, thinly sliced; reserve about ¼ cup celery leaves

Kosher salt

1 28-ounce (793 g) can whole plum tomatoes, drained and finely chopped

1 bunch Tuscan kale, stemmed, leaves torn into small pieces

Freshly ground black pepper

6 large eggs

6 slices day-old country bread

⅓ cup grated pecorino, plus more for serving

A classic recipe of the Tuscan Maremma born from peasant thriftiness, this dish reflects a barren winter landscape that offered very little for the kitchen table. Its few simple ingredients make a nutritious and satisfying meal that's still a popular mainstay on menus throughout the area. The runny yolk from the poached egg adds a nice silky texture to the soup as it mingles with the vegetables.

1. Heat the oil in a 6-quart pot over medium-low heat. Add the onions, celery, and a pinch of salt; cook until tender and just beginning to brown, 10 to 12 minutes. Add the tomatoes and cook until they begin to soften, 4 to 5 minutes. Add 6 cups water along with the celery leaves and 1 teaspoon of salt. Bring to a boil, reduce the heat to maintain a very gentle simmer, and cook for 30 minutes. Add the kale and simmer until the kale is wilted, about 5 minutes. Season to taste with salt and pepper.

2. Gently break the eggs into the soup around the edge of the pot, which will help keep the eggs together. Simmer until the eggs are just set, about 3 minutes; the yolks should still be runny. Transfer the eggs with a slotted spoon to a plate lined with paper towels.

3. Position a rack 6 inches away from the broiler and heat the broiler to high.

4. Put the bread on a baking sheet and broil until lightly golden, 1 to 2 minutes. Sprinkle with the grated pecorino and broil until the cheese is melted, about 1 minute.

5. To serve, put a slice of bread into each of six shallow bowls. Ladle in the soup over the bread. Top each bowl with a poached egg. Drizzle with a little oil and serve with grated pecorino on the side.

Tuscan black kale

SERVES 6

Maremmana Cattle and Horses

The vast, wide-open plains of the Maremma that extend down to the Tyrrhenian Sea are home to cowboys on horseback herding the long-horned cattle of the region. This present-day image may seem iconic, but the history behind it paints a much different picture of the region. Until the 1950s it was a malarial swamp. Known as butteri, Maremma cowboys lived a hard life working the disease-ravaged land. Over time, they developed immunity to malaria. Special cattle were introduced to the area in the 1600s from Hungary—the beasts' thick skin fended off disease-carrying insects—as was a big and sturdy breed of horses to help herd the giant animals. The meat from the cattle was discovered to be flavorful and lean from the animals trudging through the soggy plains, and a delicacy was born. Today the swamps have been drained and malaria eradicated, but Maremmana cows and horses stand as symbols of a once unforgiving terrain.

Broccoli Tortelli with White Sausage Ragù

TORTELLI DI BROCCOLI AL RAGÙ BIANCO

Kosher salt

8 ounces (226 g) broccoli crowns,
cut into 1-inch florets

¼ cup (59 ml) extra-virgin olive oil

1 clove garlic, sliced

Pinch of crushed red pepper flakes

1 cup (240 ml) dry white wine

12 ounces (340 g) ricotta cheese

¾ cup finely grated pecorino

½ cup finely grated Parmigiano
Reggiano

Freshly ground black pepper

1 recipe egg pasta dough (see
"Basic Recipes")

3 leeks, white and light green parts
only, cleaned and thinly sliced

12 ounces (340 g) sweet Italian
sausage, casing removed

Tortelli are a rectangular Tuscan stuffed ravioli, made from spinach, swiss chard, or even potato. The key to making the filling involves slowly cooking the vegetables until they have completely broken down. At Terra Etrusca, they make their tortelli with broccoli and ricotta, which pairs nicely with a white wine sausage sauce.

1. Bring a medium pot of well-salted water to a boil over high heat. Add the broccoli and cook until just tender, 2 to 3 minutes. Drain well.

2. Heat 2 tablespoons of the oil in a 12-inch skillet over medium heat. Add the garlic and cook until fragrant, about 1 minute. Add the broccoli, pepper flakes, and 3 tablespoons of water and cook, stirring occasionally, until the broccoli begins to break down, 12 to 15 minutes. Pour in ½ cup of the white wine and cook until the wine is gone, about 5 minutes. Season to taste with salt and pepper. Set aside to cool slightly. Pass the broccoli mixture through a meat grinder set with largest grinding plate. (Alternatively, pulse it a few times in a food processor.)

3. In a large bowl, mix together the broccoli, ricotta, pecorino, and Parmigiano until fully combined. Season to taste with salt and pepper. Transfer to a pastry bag fitted with a ½-inch plain tip.

4. Divide the dough into four pieces. Using a pasta machine, roll out the dough, starting at the widest setting and ending with the second-to-thinnest setting. Lay out one sheet of pasta and pipe tablespoons of filling down one side of the pasta sheet at 2- to 3-inch intervals. Fold the pasta over the filling. Gently press out any air between the filling and the dough. With a fluted pastry cutter, cut the tortelli into rectangles and place on a lightly floured baking sheet. Finish the remaining tortelli in the same fashion.

Making tortelli

5. Heat the remaining 2 tablespoons of oil in a 12-inch skillet over medium-high heat. Add the leeks and a pinch of salt and cook until the leeks are tender and lightly browned, 4 to 5 minutes. Add the sausage and cook, breaking up any clumps with a wooden spoon, until browned, about 3 minutes. Pour in the remaining ½ cup of white wine and cook until the wine has reduced by half, 2 to 3 minutes.

6. Bring a large pot of well-salted water to a boil over high heat. Add the tortelli and cook until they begin to rise to the surface and are al dente, 2 to 4 minutes. Reserve ¼ cup of pasta water. Drain the pasta and toss with the sausage sauce, adding some of the reserved pasta water if the sauce seems dry. Transfer to a serving platter.

SERVES 6

Hunter-Style Wild Boar
CINGHIALE ALLA CACCIATORA

3 fresh bay leaves

3 sprigs rosemary

3 sprigs parsley

3 sage leaves

2 pounds (about 1 kg) wild boar
   shoulder roast, cut into 2-inch
   pieces

1 medium yellow onion, peeled
   and halved

1 carrot, peeled and halved

1 stalk celery, halved

1 bottle (750 ml) red wine

2 tablespoons (30 ml) extra-virgin
   olive oil

1 teaspoon lightly crushed juniper
   berries

Pinch of crushed red pepper flakes

2 tablespoons tomato paste

Kosher salt and freshly ground
   black pepper

Wild boar season runs from November to mid-January. In the hills near Terra Etrusca, hunters convene on the weekends in search of trophies. During the fall months the menus of local restaurants and agriturismi are heavy with wild boar dishes.

1. Make an herb bundle by tying together the bay leaves, rosemary, parsley, and sage leaves with kitchen string. In a large bowl, add the meat, onion, carrot, celery, and herb bundle. Pour in the wine, cover with plastic wrap, and refrigerate for at least 4 hours and up to 12. Remove the meat from the red wine. Strain the wine into a bowl and set aside 2 cups. Finely chop the onion, carrot, and celery and set aside.

Hunting dogs, tired after a day of work

2. In a large pot, cook the meat over medium heat until it releases its water, 4 to 5 minutes. Remove the meat from the pot with a slotted spoon and discard the water. Wipe out the pot and set it back over medium heat. Add the olive oil, boar, chopped vegetables, herb bundle, juniper berries, and red pepper flakes and cook, stirring occasionally, until just beginning to brown, 5 to 6 minutes. Raise the heat to medium-high and pour in the reserved 2 cups of wine. Vigorously simmer until the wine is almost gone, 5 to 8 minutes. Add the tomato paste, stir well to coat the meat and vegetables, and cook until the tomato paste caramelizes a bit, 2 to 3 minutes. Add enough cold water to cover by 2 inches as well as a pinch of salt. Bring to a boil, reduce the heat to maintain a simmer, and cook until the meat is fork-tender and the liquid has reduced to the same level as the meat, 2½ to 3 hours. Season to taste with salt and pepper and serve.

SERVES 4

Tuscan hunter at lodge

Apricot Jam Roulade with Maraschino

TORTA ARROTOLATA ALL'ALBICOCCA CON MARASCHINO

1 recipe sponge cake (see "Basic Recipes")

3 tablespoons granulated sugar

2 tablespoons (30 ml) maraschino liqueur

2 13-ounce jars apricot jam

¼ cup powdered sugar

Fresh whipped cream

Maraschino, a cherry liquor, soaks into light airy sponge cake, which is then wrapped around a sweet-tart filling of floral apricot jam. Once sliced, the pinwheel design is a soft palette of orange hues.

1. Spread a clean piece of parchment paper (at least as big as the cake pan) on the counter. Sprinkle the parchment all over with sugar (this will keep the cake from sticking to the towel as it cools). Immediately after taking the cake from the oven, run a small knife around the inside edge to loosen it from the pan. Invert the cake pan onto the parchment in one quick motion. Remove the pan. Carefully peel off the parchment from the bottom of the cake. Using both hands and starting from one of the long ends, roll up the cake and the parchment together. Let cool to room temperature.

2. In a small bowl, mix together the maraschino with 1 tablespoon water.

3. Carefully unroll the cooled, parchment-wrapped cake. Brush the cake with the maraschino mixture and then spread the jam over the cake, covering it evenly to within 1 inch of the edge. Reroll the cake, without the parchment. Using two large spatulas, transfer the cake to a serving platter. Dust the cake with the powdered sugar. Cut into slices and serve with a dollop of whipped cream.

SERVES 16

Terra Etrusca's farm market

Cinnamon and Hazelnut Cookies

TOZZETTI

In the Terra Etrusca dining room, large glass jars filled with different types of biscotti sit on a rustic antique cupboard. Our favorite were these crisp crunchy cookies full of sweet cinnamon flavor, especially when dunked into glasses of Vin Santo—a Tuscan dessert wine.

1. Position a rack in the center of the oven and heat the oven to 325°F (170°C). Line two baking sheets with parchment paper.

2. In a stand mixer fitted with a paddle attachment, mix together the eggs, flour, sugar, butter, Sambucca, cinnamon, and salt until a dough begins to form. Add the hazelnuts and mix until combined.

3. Pinch off about 1 tablespoon of dough and roll out into a 3-inch log, then transfer to the sheet pan. Continue with the remaining dough. Bake the cookies one baking sheet at a time until the edges are golden brown, 30 to 35 minutes. Let the cookies cool on the baking sheet for 5 minutes and then transfer them to a rack to cool completely.

MAKES 75 COOKIES

3 large eggs

4 cups (500 g) all-purpose flour

2½ cups granulated sugar

¼ cup (½ stick) unsalted butter

½ cup (118 ml) Sambucca

2 teaspoons cinnamon

Pinch of table salt

1 cup hazelnuts, toasted and skinned

Cinnamon and Hazelnut Tozzetti with Vin Santo

FATTORIA SAN MARTINO

As their frenetic high-stress fashion careers in Milan began to wear thin, husband Antonio and wife Karin began to do some soul searching about the direction of their lives. Hoping to live a simpler life in the country, they ended up in the hills of Chianti, where they helped open and establish a winery and agriturismo. Here they became a part of a new generation of Tuscans who were attracted to the beauty and bucolic lifestyle of the countryside. Antonio and Karin were a part of a mini migration of expatriates who flocked to the region and shared a similar vision for living off the land. Through the agriturismo system, this new generation of farmers could sustain themselves by catering to the growing number of tourists seeking rural vacations in the Tuscan countryside.

Karin, a devout vegetarian since childhood, converted Antonio when they began dating, and established the farm as completely organic, learning the nuances of growing healthy crops without pesticides. They soon became interested in biodynamic agriculture and the cyclical lunar calendar and headed north for two years to study at a school in Trieste famous for this ancient philosophy. Returning to Tuscany in 2000, the couple purchased a crumbling farmhouse in the shadows of the medieval hill town of Montepulciano, and set about restoring it entirely themselves, with the intentions of opening their own agriturismo. The project continued for years, and followed the rules of bio-architecture, which call for only natural and ecological material to harmonize with the original structure. Antonio built all the beds out of reclaimed lumber, and Karin hand-sewed all of the bed linens and pillows for each room. Their eclectic, artistic flair gives the six rooms individual personality, each possessing a feeling of a modern boutique inn. Wide-planked mahogany floors, white-painted beamed ceilings, overstuffed pillows, mismatched furniture, handmade lamp shades fashioned from cloth, wire, and wood, and an infusion of contemporary and traditional art make for a tasteful and comfortable environment that feels right at home in Tuscany. Positioned in one of the most traveled corners of the region, San Martino attracts a like-minded clientele seeking a holiday not for the masses. Forgoing meat and air-conditioning were choices made by the couple to attract vacationers looking to reconnect with the spirit and natural beauty of the countryside.

At one long wooden table and in a room that is stylish, rustic, and homey, with chandeliers, a huge wood-fired stove, and antique country furniture, guests sit together to enjoy Karin's Tuscan-inspired modern vegetarian cooking. A native of Holland, Karin spent a lot of time in the kitchens of local women over the years, learning the art of making fresh pasta and local dishes. With this practical knowledge of traditional dishes, she creates her own cuisine that is strictly vegetarian, entirely organic, and extremely seasonal. A firm believer in cooking and eating for health, she disdains long-simmered sauces and lengthy cooking times that break down vegetables' nutrients. San Martino's own olive oil, honey, and jams give depth and balance to her dishes. Many are inventive seasonal salads and pastas made from San Martino's own grains, and sauced with a quick sauté of seasonal vegetables to awaken natural flavors. Muddled capers and black olives bathed in olive oil is a secret condiment, used to brighten up dishes. Karin's artistic flair shows through in her colorful presentations, right at home in her diverse dining room. In their twenty years of Tuscan living, the two have developed a network of farming friends who share similar philosophies of how they work the land, and they source ingredients from select farms around the region committed to the same principles followed at San Martino. From her own sourdough starter, Karin makes delicious bread that includes salt, defying the traditional saltless and tasteless bread of the region. Desserts are health-conscious with homemade jellies replacing sugar, and are often loaded with dried fruit and nuts for a rich and satisfying sweet. With hearty grains, a variety of nuts, good local cheeses, and organically grown produce, meat is hardly missed on San Martino's table—a refreshing alternative to the meat-driven country cooking of Tuscany.

Dining room at San Martino

Beet, Caramelized Onion, and Goat Cheese Bruschetta

BRUSCHETTA CON BARBABIETOLA, CIPOLLA E FORMAGGIO DI CAPRA

3 small yellow beets (about 12 oz./340 g), trimmed and washed

3½ tablespoons extra-virgin olive oil

Kosher salt and freshly ground black pepper

1 teaspoon orange zest

2 scallions, very thinly sliced

1 medium red onion, halved and thinly sliced

3 ounces (85 g) soft goat cheese

½ tablespoon (7 ml) whole milk

6 ½-inch-thick slices ciabatta bread, toasted

San Martino has sought to source ingredients from farmers who share similar philosophies regarding sustainability and organic growing. For their goat cheese, they traverse the windy roads of Tuscany to the Santa Margherita dairy outside Siena, where a small herd of goats are milked by hand each day to produce exceptional cheese.

1. Position a rack in the center of the oven and heat the oven to 425°F (220°C).

2. In a bowl, toss the beets with 1 tablespoon of the oil and season with salt and pepper. Transfer to a large square of heavy-duty aluminum foil and seal into a packet. Roast the beets in the oven until completely tender when pierced with a paring knife, 50 minutes to 1 hour. Let the beets cool slightly. Remove the beets from the foil, peel them, and cut them into fine dice. Toss with ½ tablespoon of the oil, along with the orange zest and scallions. Season to taste with salt and pepper.

3. Heat the remaining 2 tablespoons of oil in a 10-inch skillet over medium-low heat. Add the onion and a generous pinch of salt and cook, stirring often, until the onion is very tender and browned, 18 to 20 minutes. Transfer to a bowl.

4. In a small bowl, mix together the goat cheese with the milk until creamy. Lightly spread the goat cheese over each slice of bread. Top with the caramelized onions and then add the beets. Serve.

SERVES 6

Fattoria San Martino farmhouse

Red Beet Carpaccio with Pickled Onions
CARPACCIO DI BARBABIETOLA E CIPOLLE

This is a nice refreshing appetizer to awaken the taste buds in the wintertime. Karin uses a cheese slicer to cut her beets paper-thin, but a mandoline or a steady hand and a sharp knife will also do the trick.

1. Position a rack in the center of the oven and heat the oven to 425°F (220°C).

2. In a bowl, toss the beets with 1 tablespoon of the oil and season with salt and pepper. Transfer to a large square of heavy-duty aluminum foil, add the rosemary, and seal into a packet. Roast the beets in the oven until completely tender when pierced with a paring knife, 50 minutes to 1 hour. Let the beets cool slightly. Remove the beets from the foil and remove the peel. Using a mandoline or sharp knife, thinly slice the beets.

3. In a small bowl, combine the red onion with the red wine vinegar, sugar, and ¼ teaspoon of salt; let sit until the onion has mellowed and turned a light pink, about 15 minutes. Strain the onion and discard the liquid.

4. Using a sharp knife, trim off the peel and pith from the oranges. Thinly slice them crosswise and remove any seeds.

5. Put the garlic in a mortar with a generous pinch of salt. Use the pestle to gently grind until it forms a paste. Add the olives and capers and continue to mash into a paste. Using the pestle, stir in the remaining oil until fully combined.

6. To serve, divide the beets among six appetizer plates. Season lightly with salt and pepper. Top each plate with two orange slices, and 1 tablespoon of the red onions. Sprinkle with a few peppercorns and then drizzle with the olive pesto. Garnish with the mint leaves and serve.

SERVES 6

4 red beets, trimmed and washed

¼ cup (59 ml) extra-virgin olive oil

Kosher salt and freshly ground pepper

1 sprig rosemary

½ small red onion, thinly sliced

1½ tablespoons (22 ml) red wine vinegar

½ teaspoon granulated sugar

2 large navel oranges

½ clove garlic

¼ cup pitted small black olives

2 teaspoons salted capers

1 teaspoon green peppercorns, crushed

2 teaspoons thinly sliced fresh mint leaves

Beet Carpaccio with Pickled Onions

Cugusi

The slightly pungent bite of aged pecorino Toscano is a fairly recent addition to Tuscany's culinary heritage. After World War II, farms throughout the region were being abandoned as farmers transferred to the cities where there was a chance of better-paying work. Farmhouses were left to crumble, and fields became overgrown with brambles and weeds. On the island of Sardegna, the opposite was occurring: With an overpopulation of shepherds and not enough land to accommodate all of them, families were waging violent fights over land rights. Throughout the 1950s and 1960s, there was a migration from the island to the mainland, as a few desperate shepherds left their homes and headed to the Tuscan countryside. Here they felt at home among green rolling hills similar to the interior of Sardegna, and bought large plots of farmland for cheap.

The Cugusi family were a part of the first wave of brave souls who ventured off their island's shores. They created a new name for their family dairy with the purchase of a small parcel of land on the outskirts of Pienza. Over time they bought more and more land, built up the working farm of 1,000 sheep, and developed into one of the area's most important cheese producers. In addition to aged Pecorino Toscana, the Cugusis make a variety of sheep's cheeses, including wheels coated in tomato paste, wrapped in walnut leaves, infused with hot chile peppers, and speckled with black truffles. The Sardinian resettlement brought pecorino into the culinary repertoire of Tuscany, and today families like the Cugusi's are some of the largest landowners throughout the region. Their life-changing decision and small monetary investment have paid enormous dividends.

Sunchoke, Potato, and Onion Frittata

FRITTATA CON TOPINAMBUR, PATATE, E CIPOLLE

The sunchoke is a knobby root vegetable that sprouts a beautiful yellow flower from where it grows beneath the earth. Often used raw in salads to give a nutty, earthy crunch, it also becomes meltingly tender when cooked. Thinly slicing sunchokes helps them infuse their delicate artichoke-like flavor into this frittata as it cooks.

1. Position a rack in the center of the oven and heat the oven to 375°F (190°C).

2. In a 10-inch ovenproof nonstick skillet, heat 2 tablespoons of the oil over medium heat. Add the shallots and a pinch of salt and cook until tender, 3 to 4 minutes. Add the potatoes and 2 tablespoons of water and cook until just tender, 8 to 10 minutes. Add the sunchokes with 2 more tablespoons of water and cook until barely tender, about 5 minutes. Raise the heat to medium-high and lightly brown the vegetables, 2 to 3 minutes.

3. In a large bowl beat the eggs with the orange juice, ½ teaspoon of salt, and a few grinds of pepper. Mix in the potato mixture until fully combined.

4. Heat the remaining 2 tablespoons of oil in the skillet over medium heat. Add the egg mixture and use a wooden spoon to distribute the ingredients evenly. Reduce the heat to medium-low and cook until the eggs have set just along the outside edge of the pan, 4 to 5 minutes. Transfer the skillet to the oven and bake until the frittata is puffed, lightly golden, and set, 10 to 12 minutes.

5. Let the frittata cool in the pan for 10 minutes. Slide the frittata out of the pan and onto a cutting board. Let it cool to room temperature and then cut into wedges and serve.

SERVES 4–6

¼ cup (59 ml) extra-virgin olive oil

2 shallots, thinly sliced

Kosher salt and freshly ground black pepper

1 medium Yukon Gold potato, peeled and sliced ⅛ inch thick

6 small sunchokes, very thinly sliced (¹⁄₁₆ inch)

6 large eggs

2 tablespoons (30 ml) orange juice

Bedroom at San Martino

Roasted Potatoes and Butternut Squash with Olives and Capers

PATATE E ZUCCA AL FORNO

This starchy side complements any roasted meats, especially chicken. In her vegetarian kitchen, Karin likes to spice things up with non-Italian ingredients and here, crushed coriander seeds infuse the vegetables with their spicy sweet taste.

1. Position a rack in the center of the oven and heat the oven to 425°F (220°C).

2. In a large bowl, toss together the potato, squash, garlic, olives, capers, and coriander. Add the olive oil and mix well to combine. Season with ¼ teaspoon of salt and a few grinds of pepper. Transfer to a 9 × 12-inch baking dish and bake in the oven, stirring occasionally, until the vegetables are tender and lightly browned, 55 minutes to 1 hour.

SERVES 4

1 large Yukon Gold potato, peeled
and cut into 1-inch pieces

1 small butternut squash (about
2 pounds/900 g), peeled, seeded,
and cut into 1-inch dice (to yield
about 4 cups)

2 cloves garlic, thinly sliced

⅓ cup pitted small black olives,
such as niçoise

2 teaspoons salted capers,
rinsed well

1 teaspoon crushed coriander seeds

3 tablespoons (44 ml) extra-virgin
olive oil

Kosher salt and freshly ground
black pepper

UMBRIA

Gubbio

Villa Dama

Tevere River

Perugia

Lago
Trasmierro

Assisi

Malvarina

Spello

Fonte Antica

Todi

River Nera

Norcia

PARCO
SIBILLINI

Spoleto

Orvieto

TOSCANA

LE MARCHE

LAZIO

N

0   25   50 KILOMETERS

0   25   50 MILES

Chapter 3
⚮ UMBRIA ⚮

VILLA DAMA

Franco always dreamed of owning a house in the country. An accountant by trade in the Umbrian city of Perugia, he wanted a weekend escape from his mundane urban life. On a hunting trip with friends in the hills surrounding the medieval hamlet of Gubbio, Franco stumbled upon a deserted farm, and became enamored with its tranquil location and rugged wildness. A week later he brought his wife, Luciana, to visit the property. After seeing the state of the crumbling farmhouses and land overtaken with thick vegetation, she thought it an absurd idea. But Franco had great vision and determination and purchased the entire 370 acres.

The task before them was monumental, and would change their lives forever. Weekends were spent with friends and family who would come lend a hand with the arduous task of restoring the main farmhouse. Franco immediately cleared away a small part of the land and built a chicken coop, resurrecting the land's agricultural past, signifying his hope for the future and his family's newfound endeavor. After five years their work on the main house was finally complete, and the dream of having a place in the country a reality. Vegetable gardens, fruit trees, rabbits, and guinea fowl were introduced to the slowly expanding farm, and while the couple still resided in Perugia with their two children Danielle and

Marta, more and more time was spent at the country residence. During this time Danielle was diagnosed with leukemia. His ultimate tragic death eventually led to Franco's decision to leave his city life behind and turn his farm into an agriturismo, needing a distraction from his mourning. In 1994 the once ruined landscape of decaying rubble and overgrown terrain was inaugurated as an official agriturismo, named Villa Dama from the first two letters of *Danielle* and *Marta*.

Since opening, Villa Dama has developed into an astounding complex. From its early inception as a farm raising a few chickens, it has evolved into a completely

Pheasant

Gathering olives

organic operation of impressive variety. Aside from vegetable gardens, fields of wheat, vineyards, olive groves, bee huts, and fruit trees, the farm's main product is meat. In addition to sheep, ducks, wild game birds, rabbit, black free-range pigs, and the prized snow-white cattle known as Chianina, the farm also has a hunting reserve where wild boar and deer live freely, deep in the woods. Also on the property are an antique oil press for the fall harvest, an in-house butchery, and a mill for processing grain into flour. Using products grown, raised, and processed on the farm, Marta's cooking reflects a rustic Umbrian cuisine, full of regional character. Flatbreads baked on terra-cotta plates over burning coals served with house charcuterie, long-simmered beans with pork skin, strozzapreti pasta with shaved black truffles, thick T-bone steaks grilled over a wood fire, braised pheasant with white wine and olives, juicy pork chops, and rosemary lemon chicken roasted in the brick oven are all testimony to the impressive operations under way at Villa Dama.

The remaining buildings around the property have all been restored into apartments to accommodate a total of sixty guests, with two swimming pools built entirely of stone from the surrounding hills. Villa Dama's sprawling territory prompted Franco to purchase an old army truck that he uses to take visitors on tours of the farm down the meandering roads that snake through the property. Each year the couple hosts a special event at Villa Dama called the *Sentiero Biologico* (organic trail). Guests are each given a plate, glass, flatware, and a cloth napkin, and walk around the estate, visiting different kitchens that prepare individual courses. The final leg is at the main farmhouse, which houses a restaurant that can accommodate all 300 participants at once. Here coffee and cordials are served. The popularity of this event has surged in recent years, as has the number of people discovering Villa Dama. The entire sustainable production of the farm has left deep impressions on many guests, who return often to Villa Dama, and Franco and Luciana's living dedication in honor of their son.

Hanging salami to dry

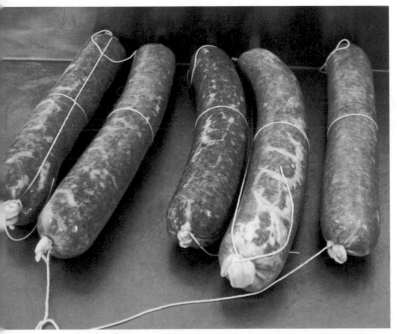

Fresh salami

Umbrian Flatbread

TORTA AL TESTO

In Umbria this traditional flatbread is cooked on a special stone over burning coals in the fireplace and served warm, cut in half, and stuffed with slices of prosciutto, pancetta, and cheese. We re-created this dish by indirectly heating a cast-iron skillet on the grill, and adding the dough to the hot pan.

3 cups (375 g) all-purpose flour

1½ teaspoons kosher salt

1 teaspoon active dry yeast

1¼ cups (300 ml) tepid water

1 tablespoon (15 ml) extra-virgin olive oil

1. Put the flour, salt, and yeast in the bowl of a stand mixer fitted with a dough hook. With the mixer running on low speed, mix in the water until fully incorporated, about 2 minutes. Raise the speed to medium and knead the dough until it comes away from the sides of the bowl and forms a ball that is smooth, elastic, and no longer sticky, 5 to 7 minutes. Oil a large bowl and add the dough. Cover loosely with plastic wrap and set aside in a warm, draft-free spot until doubled in volume, 45 minutes to 1 hour

2. Prepare a medium-high gas grill or charcoal fire. Heat a 12-inch cast-iron skillet for 10 minutes. Then prepare the grill for indirect cooking and set the skillet over the turned-off burner.

3. Turn the dough out onto a lightly floured surface and divide in half. With a rolling pin, roll out each half to an 11-inch circle about ½ inch thick. Grill one disk of dough at a time in the cast-iron skillet until air bubbles start to form, 2 to 3 minutes. With a fork, prick the dough all over and continue grilling until the dough is golden brown on the bottom, about 5 minutes. Flip the dough over and continue grilling until golden brown and dry, 5 to 10 minutes.

MAKES 2 12-INCH FLATBREADS

Baking Torta al Testo over hot coals

10 ounces (78 g) all-purpose flour

Kosher salt

1 large egg

3 tablespoons (44 ml) extra-virgin olive oil

2 cloves garlic, smashed

2 8-ounce (225 g) bunches spinach, stemmed and roughly chopped

1 6-ounce (170 g) bunch swiss chard, stemmed and roughly chopped

1 cup (250 g) ricotta cheese

Freshly ground black pepper

Spinach and Swiss Chard Roulade

FOYATA DI PERUGIA

Marta's boyfriend Simone studies at a professional culinary school in the provincial capital of Perugia. He learned this rolled bread recipe there, which has its roots in Umbria's impoverished past, and it has become a regular antipasto at Villa Dama.

1. On a clean work surface, mix together the flour and salt. Make a well in the center and add the egg, 2 tablespoons of the oil, and ½ cup water. With a fork, lightly beat the egg with the oil and water and gradually pull some of the flour into the mixture. Mix until a soft dough begins to form and then knead the dough until it becomes smooth and elastic, 6 to 8 minutes. Cover with a kitchen towel and let rest for 20 minutes.

2. Heat the remaining 1 tablespoon of oil in a large skillet over medium heat. Add the garlic and cook until golden brown, 1 to 2 minutes. Add the spinach and swiss chard, cover, and cook for 2 minutes. Uncover and continue to cook, tossing often, until just wilted, about 2 minutes more. Transfer to a colander set over a bowl, discard the garlic, and let the mixture cool. Once it's cool, squeeze out the excess water and finely chop the greens. Transfer to a bowl and mix in the ricotta until combined. Season to taste with salt and pepper.

3. On a lightly floured surface roll out the dough into a 14-inch circle, about ⅛ inch thick. Spread the mixture over the dough, covering it evenly to within 2 inches of the edges. Using both hands and starting from one of the short ends, roll up the dough; the filling may squish out of the ends a bit. Transfer seam-side down to a baking sheet lined with parchment paper. Cover loosely with plastic wrap and refrigerate for at least 30 minutes before baking.

4. Position a rack in the center of the oven and heat the oven to 350°F (180°C).

5. Bake the roulade until the dough has set and is golden brown, about 45 to 50 minutes. Transfer to a rack and let cool for at least 15 minutes. Slice into ½-inch-thick pieces and serve.

SERVES 10–12

Goat with sheep

Stewed Cranberry Beans with Pork Skin

FAGIOLI CON LE COTICHE

2 ounces (60 g) pork skin

8 ounces (226 g) dried cranberry
beans (or substitute kidney
beans), soaked overnight in cold
water

1 tablespoon (15 ml) extra-virgin
olive oil

1 small yellow onion, cut into
fine dice

1 carrot, peeled and cut into
fine dice

1 stalk celery, cut into fine dice

½ cup (118 ml) white wine

¾ cup crushed tomatoes

1 bay leaf

Kosher salt and freshly ground black
pepper

This poor man's dish has its roots with the thrifty Italian farmer, who would waste no part of the animal when it came time to feed his family. It's eaten for its heartiness and ability to satisfy with very few ingredients. The strips of boiled pork skin add an incredible depth of flavor to the beans that is filling and full of rustic character.

1. Put the pork skin in a 4-quart pot and cover with 1 inch of cold water. Bring to a boil over medium-high heat, reduce to maintain a gentle simmer, and cook until softened, about 2 hours. Drain the pork skin, rinse with cold water, and then slice into ¼-inch strips. Set aside.

2. Put the beans in a 4-quart pot and cover with ½ inch of water. Bring to a boil over medium-high heat. Reduce the heat to maintain a gentle simmer and cook until just tender, 20 to 25 minutes. Drain the beans and set aside.

3. Heat the oil in a 12-inch skillet over medium heat. Add the onion, carrot, and celery and cook, stirring occasionally, until tender and beginning to brown, about 6 minutes. Add the pork skin and cook until heated through, about 1 minute. Add the white wine to the pan and let it reduce until it's almost gone. Stir in the crushed tomatoes, bay leaf, and ½ cup of water. Simmer until the sauce begins to thicken, 8 to 10 minutes. Add the beans, and cook until the beans are creamy and very tender but still retain their shape, 15 to 20 minutes. Season to taste with salt and pepper. Serve immediately.

SERVES 6

Stewed Cranberry Beans with Pork Skin

Strozzapreti with Black Truffles and Sautéed Mushrooms

STROZZAPRETI CON TARTUFO NERO E FUNGHI

12 ounces (340 g) strozzapreti or gemelli pasta

¼ cup (59 ml) extra-virgin olive oil

8 ounces (226 g) hen of the wood mushrooms, trimmed and finely chopped

Kosher salt

2 cloves garlic, smashed

2 small black truffles, cleaned and minced

While truffles grow throughout many regions of Italy, Umbria is recognized as possessing the best and most precious of the black variety. Packs of specially trained dogs unearth the knobby spore during fall months, and the truffle decorates the menus throughout Umbria. Little manipulation is needed to bring out its flavors. At Villa Dama, this simple pasta sauce lets the delicate, earthy truffle carry the dish.

1. Bring a large pot of well-salted water to a boil over medium-high heat. Cook the pasta according to package directions until al dente. Reserve ¼ cup of the cooking water and then drain the pasta well.

2. Heat 2 tablespoons of the oil in a 10-inch skillet over medium-high heat. Add the hen of the woods and a generous pinch of salt; cook until the mushrooms have released their liquid and begin to brown, about 8 minutes. Transfer to a bowl and set aside. Heat the remaining 2 tablespoons of oil over medium heat with the smashed garlic and a generous pinch of salt. Cook until the oil is fragrant, 1 to 2 minutes. Discard the garlic. Remove the pan from the heat and let the oil cool slightly. Add the truffles to the oil and swirl to coat.

3. In a large bowl, toss the pasta together with the sautéed mushrooms and the truffle-infused oil. Toss well to combine. If the mixture seems dry, add 1 tablespoon of the reserved water at a time. Serve in individual shallow bowls.

SERVES 4

Braised Pheasant with Capers, Olives, and Sage

FAGIANO ALLA GHIOTTA

Pheasants join the ranks of the numerous animals raised at Villa Dama for the restaurant. This simple braise subdues the slight gaminess of the meat with the addition of salty olives and capers.

Season the pheasant all over with salt and pepper. Heat the oil in a 12-inch skillet over medium-high heat. Add the pheasant to the pan, in batches if necessary, and brown all over, 6 to 8 minutes total. Transfer to a plate. Add the shallot, celery, sage leaves, and capers; cook until tender and just beginning to brown, 2 to 3 minutes. Add the wine, scraping up any brown bits from the bottom of the pan with a wooden spoon, and reduce by half, 3 to 5 minutes. Return the pheasant, skin-side up, to the pan and add ½ cup of water. Bring to a boil, cover, and reduce the heat. Cook until the pheasant is tender and registers 165°F (74°C) on a meat thermometer, 15 to 20 minutes. Stir in the olives and cook until warmed through, 1 to 2 minutes. Transfer to a platter and serve.

SERVES 4

1 2½-pound (1.14 kg) pheasant, cut into 8 pieces

Kosher salt and freshly ground black pepper

2 tablespoons (30 ml) extra-virgin olive oil

1 shallot, finely chopped

1 stalk celery, cut into fine dice

16 small sage leaves

2 teaspoons salted capers, rinsed well and patted dry

½ cup (118 ml) white wine

¼ cup pitted small black olives

¼ pitted green olives

Chocolate Tart

CROSTATA DI CIOCCOLATO

½ cup (1 stick) unsalted butter, softened

⅓ cup superfine sugar

1 large egg

1¾ cups (225 g) all-purpose flour

Pinch of table salt

1 teaspoon vanilla

8 ounces (226 g) good-quality dark chocolate, such as Callebaut, chopped

¼ cup (59 ml) brewed coffee

¾ cup (175 ml) whole milk

Perugia, half an hour away from Villa Dama, lays claim to producing some of Italy's finest chocolates. One of the country's most famous industrial confectioners, Perugina, has its headquarters there, as do many boutique and artisan chocolatiers. This insanely rich and dense chocolate tart was on offer at breakfast, fueling each morning at Villa Dama with chocolate satisfaction, but would also make for a decadent dessert.

1. In a stand mixer with a paddle attachment, cream together the butter and sugar on medium speed until light and fluffy, about 3 minutes. Add the egg and mix until combined. Reduce the speed to low and mix in the flour, salt, and vanilla until a dough forms. Turn the dough out onto a work surface dusted lightly with flour and knead until smooth. Flatten the dough into a disk and wrap in plastic wrap. Refrigerate the dough for 1 hour.

2. Position a rack in the center of the oven and heat the oven to 350°F (180°C). Butter a 9-inch springform pan.

3. Put the chocolate, coffee, and milk in a metal bowl set in a skillet of barely simmering water, or in the top of a double boiler. Stir frequently with a rubber spatula until the mixture is melted and smooth. Remove the bowl from the water or double boiler and set aside to cool to room temperature, about 30 minutes.

4. Divide the dough in half. On a lightly floured surface, roll out half of the dough to a 10-inch round about ¼ inch thick; place into the prepared pan. With an offset spatula, evenly spread the chocolate over the pastry crust.

5. On a lightly floured surface, roll out the second half of dough to a 10-inch round, about ⅛ inch thick. With a fluted pastry cuter, cut eight strips about ¼ inch thick. Lay four of the strips diagonally across the chocolate filling. Lay the four remaining strips across in the opposite direction.

6. Bake in the oven until the tart is lightly golden brown and the center jiggles, 30 to 35 minutes. Set on a rack and let cool completely. Unclasp and remove the side of the springform pan. Run a long thin metal knife under the bottom crust of the tart. Carefully slide the tart onto a flat serving plate.

MALVARINA

After two decades of catering to sophisticated vacationers seeking an alternative type of holiday in the Assisi area, Malvarina has earned a distinguished reputation for its rustic character and for its preservation of the area's traditional dishes. When he followed his father's footsteps into the banking world, Claudio Fabrizi seemed destined to have a typical, comfortable white-collar career, but his eventual distaste for monotonous office work spurred a life-changing decision. With his mother, Maria, and wife, Patrizia, he opened the doors of a small working farm to overnight visitors. They began slowly with only a few bedrooms, offering just breakfast, but quickly realized the demand for serving dinner to the many guests who wanted to eat on the farm. In the tiny kitchen designed for making coffee and light snacks, Maria began cooking dishes native to the Assisi area while Claudio made guests feel at home by sitting down at the table and eating with them. Word quickly spread about this unique type of farm vacation that offered home-cooked meals by a real Italian mama, and Claudio looked to expand without losing the familiar charm of Malvarina. More bedrooms were built around the property, tucked into the hilly landscape, and a larger dining room was constructed to accommodate more guests. As they grew, friendly local women and a neighboring farmer friend, Giorgio, were hired to help keep the agriturismo running smoothly.

Nowadays dense vines lush with foliage climb the buildings that have been constructed over the years, lending them a timeless Old World charm. Brightly colored yellow and orange bee huts are lined in front of the olive groves that surround the farm while geese and turkeys roam freely among them. Horses for trekking, guinea fowl, chickens, and lamb are also raised, and a forest of beech and poplar trees provides a bounty of the famous black truffle in late fall. The two dining rooms that are now open to the public showcase the fruits of the farm as well as the products of neighboring farms. Claudio has earned much respect with the success of the agriturismo, and the office doors are plastered with stickers of recommendations from guidebooks, attesting to Malvarina as a premier destination in the countryside of Assisi.

While both dining rooms offer quaint country charm, the smaller of the two is a step back to another era. Worn wide-planked floors, long wooden tables adorned with blue-checkered tablecloths, walls ornamented with collections of antique kitchen and farming tools, and an open fireplace and brick oven lend the feeling that time has stopped. The family philosophy of preserving the traditions of the Umbrian table comes to life in the cozy room. As Maria approaches ninety, Patrizia now carries the torch in the kitchen, and in her sweet and unassuming way humbly attests to learning everything she knows from her mother-in-law. In a nod to custom, everything is prepared and presented just as it was hundreds of years ago. Lard made from Malvarina's own pigs stars as the common fat, vegetables are often wild greens foraged from the slopes surrounding the farm, and snails are gathered in nearby fields on damp mornings. Rabbits, geese, and chickens are slaughtered the morning they are eaten, and meat comes from free-range animals and often include offals and organs. A butcher block in front of the fireplace acts as a stage for presenting courses theatrically to an audience of hungry guests, who sip on tasty local wine from a nearby cooperative in anticipation. Roast goose emerges sizzling from the brick oven, its blistered skin dusted with fennel pollen and rosemary that perfumes the air as the bird is carved. Small tubes of handmade cannelloni filled with ground turkey and pork emerge bubbling from the depths of the brick oven; they're scooped from the ceramic dish they were baked in, dripping with a tomato béchamel sauce. Pork chops are seasoned liberally on the table before being set directly over the coals in the fireplace, oozing sweet succulent juice that makes the fire crackle and dance with sparks. This picture-perfect ambience retains an authentic air that is as honest and genuine as Maria and Patrizia's cooking. This strict adherence to the ways of the past has only rooted the family deeper in offering an authentic taste of Umbria.

Mother Maria preparing dinner

Black Olive, Orange, and Fennel Crostini

CROSTINI CON OLIVE NERE, ARANCIA, E FINOCCHIO

At Malvarina, they serve a typical dish of cured olives with wild fennel and orange zest. Historically, the olives were prepared during the first days of frost, when they would be left outside overnight; the cold would help in the curing process. They are delicious and impossible to stop eating. Bowls of them often begin meals at the farm. We turned this classic dish into a simple crostini whose flavors mimic those we enjoyed in Umbria.

1. Position a rack 6 inches away from the broiler and heat the broiler to high.

2. In a bowl, mix together the olives, orange zest, and fennel fronds and pollen; season with a pinch of salt and several grinds of black pepper. Add ½ tablespoon of the oil and stir well to combine. Set aside to let the flavors meld.

3. Arrange the bread slices on a rimmed baking sheet and broil on both sides until golden brown, 1 to 2 minutes per side. Drizzle one side of each slice of bread with the remaining oil; top with the olive mixture and serve.

SERVES 6

¾ cup oil-cured olives, pitted and finely chopped

3 3-inch slices orange zest, cut into fine dice

¼ cup loosely packed fennel fronds

¼ teaspoon fennel pollen

Kosher salt and freshly ground black pepper

1 tablespoon (15 ml) extra-virgin olive oil

6 ½-inch-thick slices of baguette

Ground Turkey and Pork Cannelloni
CANNELLONI ALL'ASSISIANA

1 recipe egg pasta dough (see
   "Basic Recipes")

1 tablespoon (15 ml) extra-virgin
   olive oil

8 ounces (226 g) ground turkey

8 ounces (226 g) ground pork

Kosher salt and freshly ground black
   pepper

6 ounces (170 g) ricotta

1 large egg

¾ cup grated Parmigiano Reggiano

½ cup grated pecorino

⅛ teaspoon freshly grated nutmeg

Freshly ground black pepper

½ tablespoon unsalted butter,
   softened

1 recipe tomato sauce (see
   "Basic Recipes")

½ recipe béchamel (see "Basic
   Recipes")

Every family living in and around Assisi has their own version of this recipe. Maria and Patrizia's version wraps pasta sheets around a turkey-and-pork filling livened up with nutmeg and made creamy with ricotta. A tomato sauce fortified with béchamel infuses into the cannelloni as they bake, making for a luxurious first course.

1. Divide the dough into four pieces. With a pasta machine, starting with the widest setting, roll the dough out until it is ⅛ inch thick. Cut the dough into 3½ × 3-inch rectangles. Bring a large pot of salted water to a boil. Prepare a large bowl of ice water. Slip the noodles, about ten at a time, into the boiling water and cook them until they're al dente, 2 to 3 minutes. Carefully scoop the noodles out of the water with a large wire skimmer and slide them into the ice water to stop the cooking. When they're cool, layer them between clean dish towels until you're ready to assemble the cannelloni.

2. Heat the oil in a 12-inch skillet over medium-high heat. Add the ground turkey and pork and a pinch of salt, stirring to break up any clumps, until browned, 6 to 8 minutes. Transfer to a bowl and let cool. Mix in the ricotta, egg, ½ cup of the Parmigiano, ¼ cup of the pecorino, and the nutmeg. Season to taste with salt and pepper.

3. Position a rack in the center of the oven and heat the oven to 375°F (190°C).

4. Put 1 heaping tablespoon of the meat filling in the center of each pasta rectangle and then roll it up to enclose the filling. Finish with the remaining cannelloni.

5. Butter a 14½ × 10-inch baking dish. Spread one ladle-ful of the tomato sauce evenly over the bottom. In a large bowl, mix together the remaining tomato sauce and the béchamel. Arrange the cannelloni snugly together in the baking dish in a single layer. Top with several ladlefuls of the sauce and then sprinkle with the remaining cheese. Bake in the oven until the sauce is bubbly and the top is browned, 25 to 30 minutes. Let rest for 5 minutes before serving.

SERVES 10

Patrizia preparing cannelloni

Gratinéed Cardoons in Tomato Sauce
GUBI GRATINATI IN SUGO DI POMODORO

1 bunch cardoons (about 9 stalks), peeled, trimmed, and cut lengthwise into 3 x 1-inch pieces

1 recipe tomato sauce (see "Basic Recipes")

Kosher salt and freshly ground black pepper

½ cup grated Parmigiano Reggiano

Cardoons are a staple in the autumn months, when their ribbed stalks take over vegetable gardens. Their artichoke-like flavor and celery-like texture make for an interesting combination.

1. Position a rack 6 inches from the broiler. Heat the broiler to high.

2. Bring a large pot of well-salted water to a boil over medium-high heat. Prepare a bowl with ice water and set aside. Add the cardoons to the boiling water and cook until just tender, 10 to 12 minutes. Drain and immediately transfer to the ice water. Drain again and then let dry on kitchen towels.

3. Heat the tomato sauce in a 12-inch skillet over medium-high heat. Add the cardoons and cook until very tender, 5 to 8 minutes. Season to taste with salt and pepper. Transfer to a shallow 9 X 13-inch baking dish and sprinkle with the Parmigiano. Broil until the cheese is golden brown and the sauce is a bit bubbly, 2 to 3 minutes. Serve immediately.

SERVES 4

Saltless Bread

To the foreigner, the bland saltless bread eaten and adored throughout central Italy might seem best left in the breadbasket. The dense unpalatable interior yields little in way of taste and flavor, yet as with any time-honored recipe in Italy, the story behind this bread is rich and flavorful. During papal rule, the government began imposing a high tax on salt. To the common populace, this inflation in the price of an ingredient crucial for curing and preserving food for winter was the ultimate slap in the face. Looking for ways to eliminate the pricey commodity from their diets, they began baking their bread without salt, and a newfound custom was born that continues today. Because of the tax, salt sales were regulated by the government, and for years the ingredient was only available in tobacco shops. Its presence in supermarkets is a fairly recent event, and many signs for tobacco still read SALI E TABACCHI.

Malvarina's dining room

Roasted Goose with Fennel Pollen and Rosemary

OCA AL FORNO CON FINOCCHIO E ROSMARINO

1 10-pound (4.5 kg) goose

½ lemon

Kosher salt and freshly ground
   black pepper

2 tablespoons fennel pollen

2 tablespoons chopped rosemary

Coarse sea salt

1 cup (240 ml) dry red wine

Wild fennel proliferates the Umbrian countryside, and its anise flavor plays an integral role in cuisine, especially when roasted with meats. At Malvarina, Patrizia cooks the farm's own geese in the wood-fired oven that serves as the centerpiece of the agriturismo's quaint country dining room. As guests dine on appetizers and first courses, the sweet smell of roasting fennel and goose whets the appetite for this delicious roasted bird.

1. Position a rack in the bottom third of the oven and heat the oven to 425°F (220°C).

2. Rinse and pat the goose dry. Remove excess fat from around the cavity and neck, and prick the skin of the goose all over with a fork; be careful not to penetrate the meat. Rub the goose all over with the lemon half. Season all over, including the inside cavity, with salt and pepper.

3. In a small bowl, mix together the fennel pollen and rosemary. Rub the mixture all over the goose. Sprinkle with 1 teaspoon coarse salt, and transfer to a 10 × 14-inch roasting pan.

4. Add the wine to the pan and roast in the oven for 30 minutes. Reduce the temperature to 325°F (170°C) and continue roasting the goose, basting occasionally with the pan juices, until the skin is deep golden brown and a meat thermometer registers 170°F (77°C) in the thickest part of the thigh, 45 minutes to 1 hour. Transfer the goose to a carving board and let it rest for 10 minutes before carving.

SERVES 4

Free-range geese roaming in front of bee huts

Fall Harvest Dessert
ROCCIATA

This specialty dessert hails from the Assisi area of Umbria and is traditionally made in bakeries and homes at the beginning of November following the feast of the dead and the feast of saints. It's loaded with the fruits and nuts of the fall harvest and offers a satisfying treat that is just as good for breakfast.

1. In a small bowl, mix together the dried figs, raisins, and Sambuca to soften the dried fruit, about 10 minutes.

2. In a large bowl, mix together the apple, pear, quince paste, walnuts, pine nuts, 1 tablespoon of the sugar, lemon zest, and anise seeds. Add the dried fruit and mix until fully combined.

3. Position a rack in the center of the oven and heat the oven to 350°F (180°C).

4. On a lightly floured surface, roll out the dough to a 12 × 14-inch rectangle about ⅛ inch thick. Put the filling in the center of the dough and fold the dough over the filling like an envelope. Prick the top of the dough a few times with a fork, brush with the oil, and then sprinkle with the remaining ½ tablespoon of sugar. Transfer to a baking sheet lined with parchment paper and bake in the oven until puffed up and a deep golden brown, 25 to 30 minutes. Let cool completely on a wire rack. Cut into ½-inch-thick slices and serve.

SERVES 10–12

6 dried figs (about 4 oz./115 g), finely chopped

⅓ cup golden raisins

2 tablespoons (30 ml) Sambuca

1 Golden Delicious apple, peeled, cored, and cut into fine dice

1 Bosc pear, peeled, cored, and cut into fine dice

¼ cup diced quince paste

½ cup chopped walnut pieces

¼ cup pine nuts

1½ tablespoons granulated sugar

1 tablespoon grated lemon zest

1 teaspoon anise seeds

1 pound (453 g) puff pastry dough

½ tablespoon (7 ml) extra-virgin olive oil

FONTE ANTICA

While Umbria has made its mark on the tourist map for its softly rolling, vine-covered hills and perfectly maintained medieval cities, there exists a wilder part of the region tucked away in Monti Sibillini National Park. In a land made up of sparsely populated villages, snow-peaked mountains, and crystal-clear rivers that slice through the valley, this distinctive part of Umbria has become a haven for nature lovers and for aficionados of the robust cuisine of the area. In the heart of the park and the shadow of a hilltop hamlet, the Angellini-Paroli family's agriturismo is beginning to gain notoriety for its stately presence and for the unbridled passion shared by three siblings: Paolo, Elisabetta, and Giovanni.

Fonte Antica sign

Historically the large white building housed farming families that worked for the landlord. After World War II the property was abandoned, as country folk throughout Italy migrated to the cities in the hope of higher-paying jobs. The Angellini-Paroli family eventually inherited the house, used it as a getaway from their main residence in Spoleto, and kept the historical roots of the land intact, with a neighboring farmer continuing to harvest the surrounding fields of lentils, farro, and beans. A large part of Paolo, Elisabetta, and Giovanni's childhood was made up with summer retreats to the country, whose sunny days were filled with healthy outdoor living.

In the fall of 1997, a massive earthquake struck central Umbria and left the building uninhabitable, with crumbling walls and a cracked foundation. The cost to repair the farmhouse was insurmountable, and the siblings were faced with the reality that they would need to vacate the property they all loved so dearly. But after a few years, they felt there was a gaping hole in their lives without their Umbrian country home, so they decided to restore the property into an agriturismo. With financial support from the state, patience, and a bit of tenacity, the siblings saw their idea through.

After a five-year reconstruction project, the property has been restored to its original splendor, and has the feeling of an esteemed country manor. All of the original architectural elements remain, with giant stone fireplaces and enormous vaulted arches decorating the interiors, while bedrooms are furnished with restored antiques that date back to the farm's early history. Paola, Giovanni, and Elisabetta—who had no past work experience in hospitality—thought the best way to open a place of their own would be to create an environment where they would want to spend their own vacation. A billiard lounge with a foosball table, darts, and a large open-hearth fire pit are all personal touches that add charm and comfort to the country retreat. Each night a kettle of tea is set on the fireplace hearth with a platter of homemade raisin and anise seed cookies, awaiting guests in front of a warming fire. Such attention to the smaller details of catering to their guests in a gracious and elegant setting has made Fonte Antica the ultimate rural holiday destination in the core of the rugged Umbrian hinterland.

In this untamed and untouched corner of Umbria, there exists a culture of rustic good eating, whose epicenter lies six miles away from Fonte Antica in the village of Norcia. The traditional hearty fare of the Sibillini was born here, with swine at the core of the rib-sticking cuisine; residents are recognized for their deft skills at pork butchery. Fonte Antica's antipasti usually include some type of cured pork charcuterie from Norcia, and plates come piled high with paper-thin slices of prosciutto, a variety of fresh and aged salamis, and coppa, an air-dried cut taken from the neck. On a bed of Fonte Antica's own polenta served on wooden boards sits a generous portion of pork spareribs and fennel-infused sausages that simmered for hours in tomato sauce. Second courses pay more homage to the pig, and various cuts are prepared in stews and braises or are quickly seared over the fire in the dining room. As younger generations of Italians are beginning to shun meat innards and offal, these offerings also make their way out of the kitchen, and a grilled air-dried intestine is an intense eating experience and considered a delicacy in the area, symbolic of leaner times when no part of the animal went to waste. Lentils are another famous product of the Sibillini, and eaten with great abundance for their nutritional properties and high protein content. Elisabetta makes a crumbled sausage soup with the agriturismo's lentils, whose good taste comes from being grown in the clay-rich soil of the farm. From the cold waters of the Nera River that flows through the park come fattened trout that Elisabetta sources from a nearby friend, the owner of a trout farm. Raised in a natural environment of flowing water, the fillets steam in their own juices when baked in a parchment pack, which brings out the delicate flavors of the fish. Fonte Antica's immeasurable aesthetic beauty and honest home-style cooking have contributed to the great success of this agriturismo's early years, but it's the humble dedication of Paolo, Giovanni, and Elisabetta that has resurrected the farmhouse to the glory days of their childhoods.

Polenta with Pork Sparerib and Sausage Sauce

POLENTA CON SUGO DI COSTOLETTE DI MAIALE E SALSICCE

3 tablespoons (44 ml) extra-virgin olive oil

3 ½ pounds (1 ½ kg) spareribs

Kosher salt and freshly ground black pepper

1 pound (453 g) sweet Italian sausage, pricked all over with a fork

1 medium yellow onion, cut into fine dice

1 carrot, cut into fine dice

1 stalk celery, cut into fine dice

½ cup (118 ml) dry white wine

1 28-ounce can (793 g) whole plum tomatoes, crushed by hand with their juices

2 cups medium stone-ground yellow cornmeal

½ cup grated Parmigiano Reggiano, plus more for serving

1 tablespoon unsalted butter

The people of the mountainous area in and around Monti Sibillini forged a cuisine based around what could be grown in the harsh climate. Hearty grains became a staple, and polenta became a perennial favorite for its versatility and ability to provide nourishment. A traditional way of serving the yellow porridge was on a long wooden board placed in the middle of the table; family members would dig in with their forks and eat directly from the board. At Fonte Antica, smaller cutting boards come piled high with polenta topped with this delicious sauce.

1. Heat 2 tablespoons of the oil in a 6- to 8-quart heavy-duty pan or Dutch oven over medium heat. Season the spareribs all over with salt and pepper and add to the pan. In batches, brown the ribs all over, about 3 minutes per side. Transfer to a plate. Add the sausage to the pan and brown all over, 4 to 6 minutes. Transfer to the plate with the spareribs. Pour off all but a thin layer of fat from the pan.

2. Add the remaining 1 tablespoon oil, onion, carrot, and celery to the pan. Season with a generous pinch of salt. Cook, stirring often, until the vegetables are soft and lightly browned, 6 to 8 minutes. Add the wine and cook, stirring to scrape up any browned bits on the bottom of the pot, until the liquid is reduced to almost dry, 1 to 2 minutes.

3. Return the ribs and sausage (and any juices that have accumulated) to the pan. Add the tomatoes and bring to a simmer, reduce the temperature to maintain a very gentle simmer, cover, and cook for 1½ hours. Remove the lid and continue to cook until the ribs are fork-tender, 2 to 2½ hours. Season to taste with salt and pepper. Remove the sausage from the sauce and cut into 2-inch pieces. Skim off any fat from the surface.

4. Bring 2 quarts of water to a boil in a heavy-bottomed 6-quart pot over medium-high heat. Add 1 tablespoon salt and then, in a steady stream, gradually pour in the cornmeal, whisking constantly to prevent lumping. Reduce the heat so the polenta slowly bubbles and cook, stirring often with a wooden spoon, until the polenta is tender and creamy, 30 to 40 minutes. If the polenta becomes too thick in the process, add water, a little at a time, to maintain a soft consistency. Stir in the grated Parmigiano and butter and continue to stir until the cheese is incorporated, 1 to 2 minutes.

5. Pour the polenta out onto a large serving platter. Top with the spareribs and sausage and a few ladlefuls of sauce. Sprinkle with Parmigiano and serve.

SERVES 6

Lentil Soup with Sausage and Rosemary

ZUPPA DI LENTICCHI CON SALSICCIA E ROSMARINO

1½ cups French lentils, soaked overnight in cold water

Kosher salt

2 tablespoons (30 ml) extra-virgin olive oil, plus more for serving

1 link sweet Italian sausage (5–6 oz./170 g), casing removed and crumbled

1 clove garlic, smashed

2 sprigs rosemary

1 14.5-ounce (411 g) can crushed tomatoes

Pinch of crushed red pepper flakes

Fonte Antica grows its own lentils in the clay soil of Sibillini National Park, prized for producing legumes that are tiny and tender. In the plains that stretch before the hilltop town of Castellucio, known as the piana grande, fields of lentils blaze with vivid yellows, reds, and purples from the flowering plants. It is a sight to be seen, and tourists flock from all over to witness its impressive beauty.

1. Drain the lentils, rinse, and then drain again. Put the lentils in a 3-quart pot and cover by ½ inch of water (about 5 cups). Add a generous pinch of salt and bring to a boil over medium-high heat. Reduce the heat to maintain a gentle simmer and cook until just tender, about 20 minutes.

2. Heat 1 tablespoon of the olive oil in a 4- to 5-quart saucepan over medium heat. Add the sausage and cook, breaking up any clumps with a wooden spoon, until nicely browned, 4 to 5 minutes. With a slotted spoon, transfer the sausage to a paper-towel-lined plate. Add the garlic and rosemary to the pan and cook, stirring occasionally, until the garlic is golden brown, about 3 minutes. Discard the garlic. Add the tomatoes and pepper flakes and cook for 1 minute. Stir in the lentils along with the cooking liquid, adding more water if necessary.

3. Cook the lentils until the soup is dense and creamy, but the lentils still retain their shape, 35 to 40 minutes. Stir in the sausage and cook for a few minutes more. Season to taste with salt. Serve in individual bowls drizzled with a little oil.

SERVES 4

Norcino

The small walled hamlet of Norcia, located in the sparsely populated Sibillini National Park of Umbria, has a reputation as the birthplace of the best pork butchers in all of Italy. Historically, these butchers would leave home every fall and travel throughout the country, where they would work slaughtering pigs and preserving the meat. The Norcini, as they were called, were highly sought after for their skill at curing pork, which would help feed families throughout the year. Spending long winter months away from home meant a hardened, often lonely existence. Over the years many of the Norcini began settling permanently in the towns and villages they visited.

Today descendants of these Norcini are dispersed throughout the country, and continue the work of their forefathers. A "norcineria" butcher shop is certified as offering superior-quality products—treasures of the Umbrian countryside. Norcia itself has preserved and promoted its reputation as the king of pig. The economy thrives as tourist swarm to the city to visit the numerous characteristic shops offering locally cured products, fragrant from the earthy smell of hanging salami, prosciutto, and fresh sausages.

Norcino

Trout in Parchment Paper

TROTA IN CARTOCCIO

The Valnerina valley is bisected by the Nera River, whose icy cold mountain water flows down from Monti Sibillini. Boasting some of the cleanest water in Italy, the area has become known for the many trout farms on the banks of the river. Water from the running river flows into concrete holding tanks filled with trout. Fed a natural diet and raised in the pristine waters, the fish have a sweet flavor and delicate texture. They really shine when steamed in their own juices in parchment paper packets.

4 trout fillets (about 6 oz./170 g each) or 4 whole trout (about 12 oz./340 g each), gutted and scaled

1 tablespoon (15 ml) extra-virgin olive oil, plus extra for the parchment

Kosher salt and freshly ground black pepper

2 cloves garlic, thinly sliced

4 sprigs rosemary

1 tablespoon chopped oregano

1 small lemon, thinly sliced

1. Position a rack in the center of the oven and heat the oven to 400°F (200°C).

2. Cut four pieces of parchment paper into the shape of a heart four times the size of the trout fillet (one heart per fillet). Brush the parchment with olive oil. Arrange one fillet just left of the center of the heart and season with salt and pepper. Top with a few slices of garlic, a sprig of rosemary, a sprinkle of oregano, and a few lemon wheels, and then drizzle with oil. Fold the right side of the heart over the trout fillet and crimp together the two edges of the paper to seal the parchment and make a pouch. Continue with the remaining trout fillets.

3. Brush each pouch with oil, arrange on a baking sheet, and bake for 15 minutes. Remove from the oven and let the pouches rest for 3 minutes.

4. Serve each diner an individual pouch, cutting the paper open with a sharp knife.

SERVES 4

Trout in Parchment Paper

Golden Raisin and Anise Seed Cookies
CIAMBELLINA

1 teaspoon active dry yeast

3 cups (375 g) all-purpose flour

³/₄ cup granulated sugar

½ cup (118 ml) white wine

¼ cup (59 ml) canola oil

1 large egg

½ cup golden raisins

1 teaspoon anise seeds

This recipe should appeal to lovers of soft and chewy cookies. Yeast mixed into the dough makes the cookies spring in the oven, creating a pillow that is enhanced by the sweet flavor of raisins and a mellow hint of anise.

1. In a large bowl, dissolve the yeast in ⅓ cup warm water. Add ½ cup of the flour and mix until a soft dough forms. Cover the bowl with a kitchen towel and set aside in a warm, draft-free spot until doubled in volume, about 1 hour.

2. In a stand mixer fitted with a dough hook, add the dough, the remaining flour, ½ cup of the sugar, white wine, oil, and egg. Mix on medium speed until a dough begins to form. Add the raisins and anise seeds and continue mixing until fully combined. Cover with the towel and set aside in a warm draft-free spot until doubled in volume, about 1 hour.

3. Position a rack in the top and bottom thirds of the oven and heat the oven to 325°F (170°C). Line two baking sheets with parchment paper.

4. To form the cookies, cut off a walnut-size piece from the dough and roll it out to make a rope 4 inches long. Form it into a ring and pinch the ends to close it. Dip the cookies in the remaining ¼ cup of sugar and then transfer to the prepared baking sheets. Cover with a kitchen towel and set aside until they rise to almost double in volume, about 30 minutes.

5. Bake the cookies for 10 minutes, then reduce the oven temperature to 300°F (160°C) and continue baking until the cookies are lightly golden and set, 20 to 25 minutes, rotating and swapping out the cookies halfway through. Transfer to a wire rack and let cool completely.

MAKES ABOUT 75 COOKIES

Spoleto Dessert

CRESCIONDA

The siblings of Fonte Antica still live in the splendid city of Spoleto. Many of the dishes they serve reflect the cuisine of their native town, like this baked pudding that combines bittersweet chocolate and ground amaretti cookies for a soft, creamy dessert.

1. Position a rack in the center of the oven and heat the oven to 350°F (180°C).

2. In a food processor fitted with the blade attachment, finely chop the amaretti cookies and the chocolate.

3. In a 3-quart saucepan, heat the milk over medium heat until steaming. Add the amaretti mixture to the milk and stir until the mixture has dissolved. Set aside and let cool slightly.

4. In a large bowl, beat the eggs and sugar with a handheld mixer at medium speed until thickened and pale yellow, about 3 minutes. Add the flour and the cinnamon and mix until just combined. Gradually beat in the milk mixture until combined. Pour the batter into a 9 × 13-inch baking dish. Bake in the oven until the top is set and a toothpick inserted into the center comes out with a few moist clumps, 38 to 42 minutes. Let cool completely on a rack. Cut into squares and serve.

SERVES 6

8 ounces (226 g) amaretti cookies

4 ounces (115 g) bittersweet chocolate, chopped

3 cups (720 ml) whole milk

4 large eggs

5 tablespoons granulated sugar

½ tablespoon (7 g) all-purpose flour

Pinch of ground cinnamon

LE MARCHE

EMILIA
ROMAGNA

TOSCANA

UMBRIA

MARE
ADRIATICO

Pesaro

Costa della
Figura

Urbino

Ancona

Castelraimondo

Il Giardino
degli Ulivi

Mount Sibillini
National Park

Visso

Ascoli Piceno

ABRUZZO

0   25   50 KILOMETERS
0   25   50 MILES

CHAPTER 4

LE MARCHE

COSTA DELLA FIGURA

Since opening its doors in 1997, Costa della Figura has established itself as the premier destination in the area for sampling traditional Marchigiano dishes. Slow Food immediately recognized the efforts of the owners, and Costa della Figura has been a mainstay in the organization's guides to authentic eateries throughout Italy. The small family—mother Zaira, father Enrico, and their son Marco—have upheld this reputation over the years, and the friendly familiarity of the agriturismo that greeted guests on opening day remains today. The restaurant has become a Christmas destination for a group of local chefs, who man the stoves at some of Le Marche's most decorated restaurants. They meet there annually, to celebrate the holiday with friends and family and to enjoy Zaira's honest home cooking, enhanced by her sweet and humble manner. The straightforward flavors of Costa della Figura's kitchen have established a loyal clientele from all corners of the region, who journey inland to breathe in the country air and take in the magnificent scenery. Summers have guests sitting under a veranda overlooking the farm's olive trees; winters bring diners inside the wood-paneled dining room in front of a warming fire, each location set to a theatric backdrop of American jazz from Marco's extensive collection. Both the music and Zaira's genuine dishes have become fix-tures in Costa della Figura's personality. This intimate ambience makes for a special type of country retreat.

A grove of 200 olive trees keeps the family busy during harvest season and supplies them with enough oil for all of their cooking. Enrico maintains a large vegetable garden that dictates menus at the farm, and Zaira

Mother sheep

chooses to preserve her vegetables when just picked by shrink-wrapping and freezing them, rather than follow the traditional method of conserving them in olive oil or vinegar. She also forages wild mushrooms from the hills during the foggy and damp autumn months, which she dries for use throughout the year. This allows for a balanced variety of vegetables year-round, facilitating the many different antipasti that begin meals. Smothered string beans, fava beans stewed with green peas, artichokes pan-fried and coated with bread crumbs, and gratinéed squash join generous portions of the farm's own cured pork cuts, which are slightly oversalted to compensate for the typical unsalted bread of the region. In addition to the somewhat tasteless bread is a type of focaccia Enrico bakes called crescia. While he leaves all of the cooking to Zaira, Enrico takes great pride in his role as baker, which involves making dough out of rendered pork fat, giving the crumb a pleasant chewy flavor. When the bread emerges from a wood-fired brick oven that imparts a trace of smokiness, Enrico immediately cuts it into irregular pieces and piles them high in wicker baskets, which are often returned in seconds empty.

The numerous plates of antipasti are left on the table for the entire meal, for guests to return to again and again, eventually becoming side dishes for meat entrees. First courses include bean soups and a typical egg-based fresh pasta, often sauced with rosemary-scented chickpeas and presented in the pot or sauté pan in which it was cooked, lending a comfortable homey atmosphere to the dining room. Rabbit is a constant at Costa della Figura, and a classic preparation has them cut into pieces, head and all, and roasted on top of a bed of wild fennel fronds with pancetta and garlic. Marco and Enrico swear the brain to be the most delectable, nutritious part of the animal. A wine devotee, Marco has built a cellar of Italian wines from his favorite, lesser-known producers. Lining the fireplace mantel are hundreds of books on the subject, and he enjoys introducing to his guests new wines that complement his mother's cooking. Zaira constantly pokes her head out of the kitchen throughout meals, and sits with guests who have become close friends of the families over the years. Their returning presence is a testament to the honest flavors and sincere kindness of the Costa della Figura family.

Mashed Butternut Squash Gratin

ZUCCA GRATINATA

In the fall months, the garden at Costa della Figura overflows with winter squash, and they decorate their outside veranda with the many varieties of the odd-shaped vegetable. This recipe makes a good side dish to accompany winter braises, grilled meats, or Zaira's rabbit in porchetta.

1. Put the squash in a colander and generously sprinkle with salt. Put the colander over a bowl and let the squash sit for 30 minutes, tossing occasionally.

2. Position a rack in the center of the oven and heat the oven to 425°F (220°C).

3. Rinse the squash and then pat dry. Put the squash on a baking sheet and toss with 2 tablespoons of the oil and a pinch of salt. Roast in the oven until the squash is tender when pierced with a fork and browned in spots, 25 to 30 minutes. Remove from the oven and transfer to a large bowl. Mash the squash until mostly smooth. Scoop the mash into an 8 x 8-inch baking dish and spread into an even layer.

4. In a small bowl, toss together the bread crumbs, parsley, sage, pecorino, ½ teaspoon of salt, and a few grinds of pepper. Mix in the remaining ¼ cup of olive oil, and then top the squash with the mixture. Bake until the bread crumb mixture is nicely browned, about 25 minutes. Remove from the oven and let cool slightly before serving.

SERVES 6

1 butternut squash (about 3 pounds/1.36 kg), peeled, seeded, and cut into 2-inch pieces
Kosher salt
6 tablespoons (88 ml) extra-virgin olive oil
¾ cup coarse bread crumbs
2 tablespoons chopped parsley
1 tablespoon chopped sage
½ cup grated pecorino cheese
Freshly ground black pepper

Winter squash

Chickpea Soup with Quadrucci
MINESTRA DI CECI

8 ounces (226 g) dried chickpeas

Kosher salt

5 tablespoons (73 ml) extra-virgin
olive oil

1 small yellow onion, cut into
fine dice

Pinch of crushed red pepper flakes

1 quart (946.36 ml) homemade
vegetable broth or canned low-
sodium broth

1 cup canned plum tomatoes,
chopped with their juices

¼ cup quadrucci pasta

Freshly ground black pepper

3 cloves garlic

2 3-inch sprigs rosemary

3 tablespoons grated Parmigiano
Reggiano

3 tablespoons grated pecorino

Pasta with chickpeas is a dish typical of the central and southern Italian diet, which relies heavily on vegetables and legumes. We encountered versions of this dish in nearly every region, served in big bowls and bathed in glistening olive oil. The meatiness of the beans makes for a richly satisfying and filling meal that makes you think twice about ordering a second course. Dried chickpeas make all the difference in this recipe and are well worth the extra step.

1. Put the chickpeas in a large bowl and cover with water and a generous pinch of salt. Let them soak in the refrigerator overnight. Drain the chickpeas, rinse well, and then drain again.

2. Put the chickpeas in a 4-quart pot, cover with water by 1 inch, and bring to a boil over medium-high heat. Reduce the heat to maintain a simmer and cook until the chickpeas are tender and creamy, 35 to 40 minutes. Drain the chickpeas and set aside.

3. Heat 2 tablespoons of the olive oil in a large pot over medium heat until shimmering. Add the onion, crushed red pepper, and a pinch of salt and cook until tender and beginning to brown, 5 to 7 minutes. Add the chickpeas to the pot along with the vegetable broth, tomatoes, and 1 cup of water. Bring the mixture to a boil, reduce the heat, and let simmer until the flavors have melded together, 15 to 20 minutes. Add the pasta to the soup and cook until the pasta is al dente, following the instructions on the package. Season to taste with salt and pepper.

4. Meanwhile, heat the remaining 3 tablespoons of oil, with the garlic cloves and the rosemary sprigs, in a small saucepan over medium heat until the oil is fragrant, about 2 minutes. Remove from the heat.

5. Transfer the soup to a large serving bowl and sprinkle with a generous amount of the Parmigiano and pecorino cheese. Drizzle with the infused oil and then garnish with the rosemary sprigs from the oil.

SERVES 6

Le Marche landscape

Chicken Ravioli with Cinnamon and Pecorino

RAVIOLI DI GALLINA

Simmering ravioli in the same liquid used for cooking chicken and pork infuses this delicious pasta with flavor. The cinnamon adds an interesting note and the lemon a bit of brightness, allowing for these ravioli to stand on their own without a sauce.

½ chicken (about 1 pound/453 g)

6 ounces (170 g) ground pork, left in one solid piece

1 yellow onion, halved

1 carrot, peeled

1 stalk celery

Kosher salt

¾ cup bread crumbs

1½ cups finely grated pecorino

2 eggs

2 teaspoons lemon zest

¼ teaspoon cinnamon

1 recipe egg pasta dough (see "Basic Recipes")

1. Put the chicken, pork, onion, carrot, and celery in a 5-quart stockpot and add water to cover by 1 inch (about 3.5 quarts). Bring the water up to a boil over medium-high heat. Add 1 tablespoon of salt; reduce the temperature and simmer, skimming away any scum that rises to the surface, until the chicken is just cooked through, 35 to 40 minutes.

2. Remove the chicken and the pork from the broth with a slotted spoon and set aside to cool. Strain the broth through a fine-mesh sieve into a clean large pot and discard the vegetables.

3. When it's cool enough to handle, remove the meat from the chicken and shred into large clumps. Discard the skin and bones. Put the chicken and pork into a food processor and pulse until finely chopped, but not a paste. Transfer the meat to a large bowl.

4. In a small bowl, dampen the bread crumbs with 2 tablespoons of the broth. Add the bread crumbs to the meat mixture along with 1 cup of the cheese. Add the eggs and mix well. Add the lemon zest, ⅛ teaspoon of the cinnamon, and a pinch of salt and mix with a wooden spoon until the mixture is homogeneous.

5. Divide the dough into four pieces. Using a pasta machine, roll the dough out, starting at the widest setting and ending with the second-to-thinnest setting. Lay out one sheet of pasta and drop tablespoons of filling down one side of the sheet at 1-inch intervals. Fold the pasta over the filling, and gently press out any air between the filling and the dough. With a pastry cutter or knife, cut the ravioli into squares and place on a lightly floured sheet pan. Finish the remaining ravioli in the same fashion.

6. Add some water to the pot with the broth and bring up to a boil over high heat. Drop the ravioli into the pot and cook until al dente, 2 to 3 minutes. Remove the ravioli with a slotted spoon and divide among six shallow bowls. Sprinkle with the remaining cheese and cinnamon and serve.

SERVES 6

Pecorino cheese

Rabbit Cooked in Porchetta Style
CONIGLIO IN PORCHETTA

1 tablespoon fennel pollen

Kosher salt and freshly ground black
pepper

2 rabbits (about 2 pounds/1 kg
each), cut into 8 pieces

3 tablespoons (44 ml) extra-virgin
olive oil

5 ounces (141 g) pancetta,
cut into small dice

3 small garlic cloves, sliced

½ cup (118 ml) white wine

½ cup small black olives

In porchetta refers to a style of cooking that emulates porchetta, a whole roasted pig rolled, tied, and stuffed with wild fennel. At Costa della Figura, they serve the rabbit with the traditional porchetta flavorings, but instead of deboning the rabbit, stuffing it, and rolling it, they use a less tedious method and simply cut the rabbit into pieces. They also add their own twist by adding some of the black olives that they grow.

1. Position a rack in the center of the oven and heat the oven to 425°F (220°C).

2. In a small bowl, mix together the fennel pollen with 1 teaspoon salt and ½ teaspoon black pepper. Rub the mixture all over the rabbit pieces. Heat the oil in a flameproof 12 × 14½-inch roasting pan over medium-high

IN PORCHETTA

Porchetta is a classic central Italian preparation of pork that involves deboning an entire pig, heavily seasoning it with salt and pepper, garlic, rosemary, and fennel, and slow-roasting it in a wood-fired oven until the skin becomes blistered and crackly. Thinly sliced shavings of the meat make for a delicious sandwich that's widely eaten throughout Lazio, Umbria, Tuscany, Le Marche, and the Abruzzo, and sold out of trucks at markets and along roads. In Le Marche they reproduce the flavors of this classic dish with the preparation of many of their second courses. The Marchigiana in porchetta refers to dishes made with pancetta and wild fennel fronds that grow copiously throughout the region. Common variations include duck, rabbit, eggs, goose, and even seafood, all of which benefit from the harmonious combination of crispy bacon and anise.

heat. Add the rabbit and brown all over, 6 to 8 minutes. Transfer to a plate. Add the pancetta and garlic and cook until the pancetta renders its fat and browns, about 5 minutes. Add the white wine and reduce by a quarter. Add the rabbit back to the pan along with the olives and cook until the rabbit is tender and a meat thermometer registers 165°F (74°C), 25 to 30 minutes.

3. Remove from the oven and transfer to a serving platter. Serve immediately.

SERVES 6

FOSSA CHEESE

The histories behind some of Italy's most prized ingredients are often as rich and delicious as the products themselves. In Le Marche there exists a special type of pecorino cheese called Pecorino di Fossa—named aptly for a type of porous rock that is carved and made into an airtight container for aging the cheese. The tradition began in the Middle Ages when peasants began burying their precious food products underground, to protect them from being plundered by invading armies. Large irregular pieces of slightly aged pecorino were wrapped in hay and placed in hollowed-out chunks of fossa rocks. While underground, the cheese had developed a pleasant earthiness, absorbed from the dirt and soil, and a specialty was born. Today strict rules mandate the preparation of Pecorino di Fossa. The cheese gets buried at the end of every August and must remain there for ninety days, until November, when festivals across Le Marche commemorate the cheeses' unearthing.

Nutella Sandwich Cookies

PESCHE ALLA MARCHIGIANA

4 cups (500 g) all-purpose flour

1 teaspoon baking powder

1/2 teaspoon baking soda

Pinch of table salt

4 large eggs

1 cup granulated sugar

1 teaspoon lemon zest

1 teaspoon (5 ml) vanilla extract

4 1/2 tablespoons unsalted butter, melted

1/2 cup Nutella

These cookies are traditionally dipped into a red cinnamon-flavored liquor called Alchermes, dyeing them an orangey color, giving them the name pesche—Italian for "peach." We found they are just as good without the hard-to-find liquor and less time consuming to make, not to mention much cleaner; you won't have to spend time scrubbing away the red dye from your hands and fingers.

1. Position a rack in the center of the oven and heat the oven to 350°F (180°C).

2. Sift together the flour, baking powder, baking soda, and salt in a medium bowl. In a large bowl, add the eggs, sugar, lemon, and vanilla and whisk until combined. Whisk in the butter until incorporated. Stir in the flour mixture with a wooden spoon until just combined.

3. Line two to three baking sheets with parchment paper. Spoon out 1/2 tablespoon of dough, dampen your hands with water, and roll the dough into a ball. Place on the prepared baking sheets, about 1/2 inch apart, and gently press down to flatten. Continue with the remaining dough. Bake one sheet at a time until the cookies are lightly golden and feel dry to the touch, 10 to 14 minutes. Cool completely on racks.

4. Drop about 1 teaspoon of the Nutella in the center of a cookie, top with another cookie, and gently press them. Repeat with the remaining cookies. Store in an airtight container for up to 2 days.

MAKES ABOUT 24 COOKIE SANDWICHES

Nutella Sandwich Cookies

IL GIARDINO DEGLI ULIVI

A unique agricultural niche—raising trotter racehorses—and a distinctive farmhouse built into an attached village set the Cicciolini family agriturismo apart from others in the region. Husband Santo and wife Maria Pia followed their dreams of restoring Santo's family's forsaken and crumbling property and turning it into the area's first established agriturismo. An architect by profession, Santo wanted the building to retain the intrinsic elements of the original structure, and to blend into the landscape of the town and the surrounding farmland, offering both an urban and rural feeling to the property. Over five years the couple meticulously restructured the entire building, an arduous project that consumed all of their free time. During the years of construction, their son Francesco was establishing the farm as a reputable breeder for a type of Italian horses, known as trotters, whose lean, muscular bodies were suited for pulling carriages. By the time of Il Giardino degli Ulivi's grand opening, the farm enjoyed significant recognition within the European horse world, and the Cicciolini family was ready to take their agriturismo in a similar direction. In August 1991 the agriturismo opened its doors to a flood of curious guests from the area who had witnessed the farm's rebirth and rise from the rubble.

In its twenty-year history, Il Giardino degli Ulivi has amassed a slew of awards and decorations for its commitment to preserving local food cultures as well as being a charming place to stay in the Le Marche countryside. These days Maria Pia looks back with great fondness on the sacrifices she and Santo endured together in the farmhouse's reconstruction, the thousands of guests they entertained in their dining room, and the joy they found together sharing their farm and lifestyle with others. Since Santo's recent passing, the farm has entered a new phase, with Francesco and his wife, Dory, taking more of a role in the agriturismo. A large gap has been left with Santo's absence, but Maria Pia's cooking remains as genuine and honest as it was on opening day, steadying her evolving future.

Over the years Maria Pia's kitchen has earned praise for its refined rusticity in representing Marchigiano cuisine. In its early days the restaurant had its own professional chef, whom she credits with broadening her palate and taking her own cooking to greater heights. Since his departure, Maria Pia has taken over the stove, and a constant in her cooking is the love and dedication she puts into her craft. A firm believer in the pleasing aesthetics of artfully plated food, she adorns dishes with leaves from grapevines, edible flowers from the garden, and hand-carved skewers from branches of laurel bushes. The artful presentations are at home beneath the stone arches, ancient fireplace, antique tables, and painted depictions of horses that make up the dining room. Her cooking style blends tradition with influences from other regions and modern twists. A special type of yeasted pasta dough has a wonderful toothy and chewy, almost bread-like texture, and is served with the classic amatriciana sauce from nearby Lazio. The egg, bacon, and pasta combination made famous by the Romans makes frequent appearances at the farm's table, and Maria Pia's intensely silky carbonara would give any chef from the eternal city a run for its money. On a trip to Rome to visit her daughter a few years ago, she was first introduced to ginger, which she now uses as an important component in her marinades. Pork loin bathed in sweet wine infused with the root emerges with a hint of gingery flavor, enhanced when roasted over a bed of apples. Boiled-down grape juice coats pan-fried Cornish game hens, dyeing them deep purple and imparting a delicate sweetness. An impressive, well-appointed wine list joins meals, created by Santo after earning a professional sommelier degree; a favorite pastime of his was driving throughout the Italian countryside in search of lesser-known wineries. Santo developed deep personal contacts with winemakers throughout the country, which afforded him the opportunity to bring into his restaurant labels not accessible to all buyers. Since his passing, Francesco has taken over the wine buying, maintaining similar relations with area vineyards. He looks to continue his parents' hospitable success and to carry the torch of Il Giardino degli Ulivi's long run into the next generation.

Room with a view

Ricotta Custard with Lemony Shaved Asparagus

TIMBALLO DI RICOTTA CON ASPARAGI

12 ounces (340 g) ricotta

1 large egg plus 1 yolk

¼ cup plus 2 tablespoons (88 ml) heavy cream

2 tablespoons grated Parmigiano Reggiano

1 tablespoons grated pecorino

1 tablespoon grated lemon zest

2 teaspoons chopped marjoram, plus more for garnish

2 tablespoons (30 ml) extra-virgin olive oil, plus more for the ramekins

Kosher salt and freshly ground black pepper

6 ounces (170 g) asparagus, trimmed

2 teaspoons (10 ml) lemon juice

Good-quality ricotta is an essential ingredient in the Italian pantry. These soufflé-like custards have a luscious texture that is enhanced by the subtle nuances of lemon and mint. While the recipe calls for asparagus, other vegetables can be used depending on the season.

1. Position a rack in the center of the oven and heat the oven to 350°F (180°C).

2. Put the ricotta, egg, yolk, heavy cream, Parmigiano, pecorino, 2 teaspoons of the lemon zest, marjoram, 1 tablespoon of the oil, and 1 teaspoon salt in a large bowl. Whisk the ingredients until smooth and thoroughly combined. Oil five 4-ounce ramekins. Fill the ramekins about two-thirds of the way with the ricotta mixture. Put the ramekins in a 9 × 13-inch baking dish. Add enough warm water to come halfway up the sides. Bake in the oven until the mixture has puffed and is lightly golden brown, 35 to 40 minutes.

3. Meanwhile, remove the tips from the asparagus and set them aside in a large bowl. With a vegetable peeler shave the asparagus, discarding the first shavings. Add the shavings to the bowl with the tips. Mix in the remaining 1 teaspoon of lemon zest, the lemon juice, ¼ teaspoon of salt, and a few grinds of pepper. Add the remaining 1 tablespoon of oil and toss well to combine.

4. To serve, invert the ricotta custards onto five small appetizer plates. Top each with the shaved asparagus and some marjoram and serve.

SERVES 5

Cornish Game Hens with Saba
GALLETTO CON SAPA

Saba, also known as mosto cotto, is a sugary, concentrated dark syrup made from the boiling down of freshly squeezed grape juice. Believed to date back to Roman times and thought to be the predecessor to balsamic vinegar, saba has forever been a staple of the Italian pantry. At Il Giardino degli Ulivi, Maria Pia uses the syrup to create a densely concentrated sauce with the syrup that is full of herbaceous notes from the plethora of aromatics.

1. With poultry shears, remove the backbone from the Cornish game hens. Split the hens in half through the breastbone with a sharp knife. Season the birds all over with salt and set aside.

2. Heat the oil in a 12-inch skillet over medium-high heat. Add the hens skin-side down and cook until deeply golden brown, 3 to 4 minutes. Remove from the pan. Add the pancetta and cook until it has rendered some of its fat, 3 to 4 minutes. Add the sage, thyme, rosemary, mint, bay leaves, garlic, ginger, and juniper berries; cook until fragrant, about 1 minute. Add the game hens skin-side up, cover the pan, reduce the heat to low, and cook until the game hens are almost cooked through, about 20 minutes. Remove the lid, raise the heat to high, pour in the cognac, and then carefully light the cognac on fire with a long match or grill lighter. Shake the pan gently back and forth until the flames subside. Add the broth to the pan and reduce by half.

3. In a small bowl, mix together the saba and the vinegar. Add the saba mixture to the pan and cook, stirring to combine, until the flavors have melded together, 2 to 3 minutes. Season to taste with salt. Transfer the game hens to a large serving platter. Strain the sauce through a fine-mesh sieve, and then drizzle the sauce over the hens. Serve immediately.

SERVES 4

2 Cornish game hens (about 1½ pounds/680 g each)

Kosher salt

1 tablespoon (15 ml) extra-virgin olive oil

2 ounces (60 g) pancetta, cut into small dice

6 whole sage leaves

3 sprigs thyme

2 3-inch sprigs rosemary

2 small sprigs mint

2 bay leaves

2 whole cloves garlic, unpeeled

½-inch piece of ginger, thinly sliced

½ teaspoon juniper berries, lightly crushed

¼ cup (59 ml) cognac

¼ cup (59 ml) low-sodium chicken broth

2 tablespoons (30 ml) saba

1 tablespoon (15 ml) cider vinegar

Bread Dough Pasta with Amatriciana

PINCIANELLE ALL' AMATRICIANA

4 cups (500 g) all-purpose flour

1 teaspoon granulated sugar

½ teaspoon active dry yeast

Kosher salt

1 egg white

1 tablespoon (15 ml) extra-virgin olive oil

2 ounces (60 g) guanciale or pancetta, cut into fine dice

1 small yellow onion, cut into fine dice

Pinch of red pepper flakes

1 28-ounce (793 g) can whole plum tomatoes, finely chopped, with their juices

2 bay leaves

Grated pecorino cheese, for serving

Maria Pia served us steaming bowls of this dish in front of a raging fire on a blustery afternoon upon on our immediate arrival to Giardino degli Ulivi. We still talk about how good it was. Referred to as pincianelle, which means "to pinch" in dialect, this unique pasta was once made by pinching off small bits of dough and rolling them out by hand into 1-inch strips. What makes this dish so delightful is the inclusion of yeast into the dough, which gives a chewy texture to the pasta. Today these are much easier to make with the help of a pasta machine.

1. In a stand mixer fitted with a dough hook, mix together the flour, sugar, yeast, and 1 teaspoon salt on low speed. Add the egg white and 1¼ cups tepid water and mix together at medium-low, adding more water as necessary, until a soft dough forms. Raise the speed to medium and knead the dough until it is very smooth and elastic, 8 to 10 minutes. Cover the bowl with a kitchen towel and put it in a warm, draft-free spot, until the dough has almost doubled in volume, about 1 hour.

2. Turn the dough out onto a floured work surface and knead a few times. Divide the dough into six pieces, and then pass one piece through a pasta machine on the widest setting. Dust the rolled piece of dough lightly with flour and fold in half. Run it through the machine again. Repeat this step two to three more times. Rest the sheet of dough on a clean tablecloth. Repeat with the remaining pieces of dough.

3. On a generously floured surface, cut the sheets of dough lengthwise into 2-inch strips and then cut each strip crosswise into ⅛-inch-thick strips. Toss the strips generously with flour, and then transfer to a floured sheet tray. Put the sheet tray in the freezer until the pasta is completely frozen, 30 to 35 minutes.

4. Meanwhile, make the sauce. Heat the olive oil in a 4-quart saucepan over medium heat. Add the guanciale and cook until it has rendered its fat and is beginning to crisp, 4 to 5 minutes. Add the onion and pepper flakes and cook until the onion is tender, about 4 minutes. Add the tomatoes and bay leaves, bring to a simmer, and cook until the sauce is fragrant, 30 to 35 minutes. Discard the bay leaves and season to taste with salt.

5. Bring a large pot of well-salted water to a boil over medium-high heat. Drop in the pasta and cook until they begin to rise to the surface, about 3 minutes. Drain the pasta, toss with the sauce, and serve with the grated pecorino on the side.

SERVES 4 TO 6

Pincianelle all'Amatriciana

Pork Loin Roast in Sweet Wine with Apples, Ginger, and Onions

LOMBO DI MAIALE CON ZIBIBBO, MELE, E ZENZERO

Maria Pia offers Zibibbo, an amber-colored sweet wine from Sicily, as an after-dinner drink to her guests. In this dish, she marinates pork loin overnight in the wine with ginger and herbs for a delectable roast infused with floral and honey accents. Be sure to slice the loin paper-thin. Any leftovers make for great sandwiches.

1 2-pound (1 kg) boneless pork loin roast

Kosher salt

2 tablespoons (30 ml) extra-virgin olive oil

1 cup (240 ml) sweet wine, such as Zibibbo

4 3-inch sprigs rosemary

8 sage leaves

4 whole cloves garlic, unpeeled

2 medium yellow onions, cut into medium dice

2 red apples, like Braeburn or Rome, cut into medium dice

1 3-inch piece of ginger, peeled and halved

1½ tablespoons unsalted butter, cut into pieces

1. Season the pork lightly with salt. Put it in a large bowl with the olive oil, sweet wine, rosemary, sage, and garlic; turn to coat. Cover with plastic wrap and refrigerate at least 4 hours and up to 24 hours, turning every so often.

2. Position a rack in the bottom third of the oven and heat the oven to 375°F (190°C).

3. In a large bowl mix together the onions, apples, ginger, and 1/2 teaspoon salt. Remove the pork from the marinade, pour the marinade mixture in with the apples, and toss to combine. Put the apple mixture into the bottom of a 9 × 13-inch roasting pan. Put the pork on a rack and position the rack over the apple mixture.

4. Arrange the pieces of butter on top of the pork and roast in the oven until the pork is just cooked through and registers 140°F (60°C) on a meat thermometer, 25 to 30 minutes. Turn the broiler to high and broil until the top browns, 2 to 3 minutes. Remove the pork from the oven and transfer to a carving board. Tent loosely with foil and let the roast rest for 10 minutes before carving. Carve the meat into thin slices and serve with the apple mixture on the side.

SERVES 4–6

Chapter 5

Lazio

CASALE VERDELUNA

In a region not readily known for its oenologists' prowess, the Casale Verdeluna cantina in the Ciociaria section of Lazio is among a select group of winemakers beginning to gain notoriety for their production of top-notch wines. Lino Nardone has been patiently experimenting and perfecting his wines, and he now offers an exceptional line from Lazio's first DOCG, known as Cesanese. Desiring a career change from a lifetime of corporate work, Lino purchased a neglected farmhouse in the countryside and immediately planted the small plot of land with Cesanese vines. He then commenced work on the decaying stone farmhouse, whose sorry state required a near-complete demolition before construction could begin. All of the original materials were salvaged and used to piece back together the structure to create the cantina and agriturismo. After a few years of trial and error, Lino's wines are beginning to find their own personality and style. He makes several lines of Cesanese that range from a young and fruity type fermented in stainless steel, to a more rustic version with pronounced tannins from time spent in large wooden barrels, to a soft and robust bottling aged in small barrique oak barrels. As Casale Verdeluna makes a name for itself among other wine producers in Lazio, the agriturismo has followed suit in delivering a tranquil retreat easily acces-

sible from Rome. Surrounded by vines and the sounds of nature, the winery has become a favorite destination of city dwellers, who fill the dining room on weekends. Seated at rustic wooden tables beneath knotty beams and exposed brick walls, they embrace the agriturismo's familiar atmosphere. The kitchen serves up tasty plates of characteristic Roman fare.

Lino has two great local women cooks, Diana and Donata, who work together to bring an authentic Roman dining experience to Casale Verdeluna. Each has her own role, with Diana preparing the sauces and working the stove during dinner, while Donata focuses on rolling and shaping all of the pasta and baked goods. Everything is done by hand here, including the making of egg pasta, which each morning is rolled out with rolling pins, skill, and inexhaustible patience. The slight irregularities in shape and cut give each bite of pasta a distinctive texture that no machine can replicate. The same can be said for the potato gnocchi, made from potatoes gown on the farm, which are cut into thumb-size dumplings and served with a fresh tomato sauce in summer or with squash and pancetta, and local smoked provolone in colder months. Diana bakes all of the bread in a brick oven and also deep-fries focaccia made with house olive oil to accompany platters of prosciutto and just-picked figs. Lamb is a regular on the menu, satisfying the Lazio

LAZIO

TOSCANA

UMBRIA

ABRUZZO

MOLISE

CAMPANIA

MARE
TIRRENO

○ Rieti

*Le Mole
Sul Farfa* ⌂

○ Poggio Moiano

⍟ **Roma**

*Casale
Verdeluna* ⌂

0 25 50 KILOMETERS
0 25 50 MILES

N

love of the meat. It's paired with artichokes, another native ingredient, for a ragù served over penne; or skewered and grilled; or roasted with plenty of rosemary plucked from the huge bushes outside the kitchen door. Suckling pigs, another staple on the menu, are whole-roasted on beds of fennel for a delectable porchetta-style second course. Desserts round out the meal and include the classic torta della nonna filled with a generous spread of pastry cream and pine nuts; a milky ricotta, apple,

and cinnamon tart; and a fluffy crepe folded over a bed of Chantilly cream and topped with a sweet sauce of roasted grapes plucked straight from the vines. Lino's budding cantina and agriturismo has quickly evolved into a formidable representative of Lazio's gastronomy, and those willing to venture beyond Italy's name-brand wines and seek out the smaller, less recognized producers, like Casale Verdeluna, will be rewarded with exceptional drinking value.

Casale Verdeluna

Ground Lamb and Artichoke Ragù
RAGÙ D'AGNELLO E CARCIOFI

1 lemon, quartered

10 baby artichokes

3 tablespoons (44 ml) extra-virgin olive oil

Kosher salt

2 bay leaves, preferably fresh

2 sprigs thyme

1 clove garlic, minced

Pinch of crushed red pepper flakes

1½ pounds (680 g) ground lamb

2 carrots, peeled and minced

¼ teaspoon freshly ground pepper

½ cup (118 ml) dry white wine

1 pound (453 g) penne rigati

1 tablespoon unsalted butter

Throughout the countryside of Lazio you'll find fields lush with blooming arti-chokes and flocks of sheep out to pasture—two of the region's most cherished ingredients, and this recipe combines them. The finely chopped artichokes melt into the sauce, adding their delicate flavor to the ground lamb.

1. Squeeze the lemon quarters into a large bowl filled with cold water and then add to the bowl. Clean the baby artichokes by removing the dark green outer leaves until only the pale, tender inner leaves remain. Trim ¼ inch from the top of the artichokes, and then trim the stem end and any dark parts around the bottom. Cut the artichokes in half and then add them to the bowl of lemon water. Continue cleaning remaining artichokes.

2. Drain the artichokes and pat dry with kitchen towel. Put them in a food processor fitted with the blade attachment and pulse until very finely chopped. Heat 1 tablespoon of the oil in a 10-inch skillet over medium-high heat. Add the artichokes and a pinch of salt and cook until browned, 3 to 4 minutes. Set aside.

3. Heat the remaining 2 tablespoons of oil in a heavy-bottomed 6-quart pot over medium-high heat. Add the bay leaves, thyme, garlic, and crushed red pepper, and cook until the oil is fragrant, about 1 minute. Add the ground lamb, carrots, 1 teaspoon of salt, and pepper. Cook, stirring occasionally, until browned, 6 to 8 minutes. Add the white wine and reduce to almost dry, 6 to 8 minutes. Stir in the artichokes, reduce the heat to medium-low, and simmer the ragù until the flavors have melded and the artichokes are fully cooked, 10 to 12 minutes.

4. Meanwhile, bring a large pot of well-salted water to a boil over high heat. Drop in the penne and cook until al dente, according to the box instructions. Reserve ¼ cup of pasta water and then drain the pasta well.

5. Toss the pasta with the sauce. Add 2 tablespoons of the pasta water and the butter and toss until combined, adding more water 1 tablespoon at a time if the sauce seems dry.

SERVES 4

Making fresh pasta

"Little Rags" of Beef with Radicchio
STRACCETTI AL RADICCHIO

3 tablespoons (44 ml) extra-virgin
   olive oil

2 cloves garlic, smashed

1 large head radicchio, halved, cored,
   and thinly sliced

1 teaspoon minced rosemary

Kosher salt and freshly ground
   black pepper

1 pound (453 g) skirt or flank steak,
   cut crosswise into paper-thin
   slices

1 ounce (28 g) shaved
   Parmigiano Reggiano

An iconic dish of both Rome and Lazio, straccetti's simple preparation makes for a quick and easy Italian-style stir-fry. The name translates as "little rags," for the beef's resemblance to shredded cloth. This satisfying second course can either be made with radicchio or arugula. Be sure to slice the beef as thinly as possible, and be careful not to overcook the meat.

1. Heat 2 tablespoons of the oil in a large cast-iron skillet over medium heat. Add the garlic and cook until the oil is fragrant, about 1 minute. Discard the garlic. Add the radicchio, rosemary, a pinch of salt, and a few grinds of pepper; cook, stirring occasionally, until wilted and tender, 3 to 4 minutes. Transfer the radicchio to a plate.

2. Heat the remaining 1 tablespoon of oil in the pan over medium-high heat. Season the beef all over with salt and pepper and then add to the pan and cook, stirring often, until the meat loses its raw color, 1 to 2 minutes. Add the radicchio to the pan and cook for 1 minute more, to combine the flavors. Transfer to a platter and top with the shaved Parmigiano.

SERVES 4

SCARCHILLI

Rome's urban sprawl has penetrated the core of Lazio and deep into the once rural countryside outside the city walls. In the area known as Ciociaria in the province of Frosinone, most agricultural activities have given way to industry, and the landscape is dominated by commerce, infrastructure, and heavy traffic.

In the center of all of this modernity, however, the small dairy of the Scarchilli family has weathered the changing times by remaining committed to producing a special type of smoked cheese since 1933. Gran Cacio di Morolo is made via a lengthy and arduous process, which begins by turning cow's milk into cheese that is molded into gourd-like shapes known as caciocavallo. These are smoked over beech wood and hung to dry for a short time. From there they are set atop wood shavings in individual boxes and left to age from six months to a year; a lucky few rest for up to two years to achieve grand reserve status. The cheese is colored a golden yellow resembling a Dutch Gouda, and its intense flavors of smoke are offset with a pleasant infusion of fresh-cut woodiness. Lino has become good friends with the Scarchilli family over the years and supports their dedication to their craft by stocking the cheese as a staple in Casale Verdeluna's kitchen.

Red Wine Cookies

CIAMBELLINE AL VINO ROSSO

4 cups (500 g) all-purpose flour

³/₄ cup (180 ml) red wine

³/₄ cup (180 ml) extra-virgin olive oil

1¼ teaspoons baking powder

1 teaspoon anise seed

1 teaspoon baking soda

1 cup granulated sugar, plus more
    for the cookies

Traditional to Lazio, these wine cookies are made in homes and bakeries through-out the region. A good-quality wine is essential, and at Casale Verdeluna they use a robust red. These cookies are great any time of day and are quick and easy to make.

1. Position a rack in the center of the oven and heat the oven to 350°F (180°C). Line two baking sheets with parchment paper.

2. In a stand mixer fitted with the paddle attachment, add the flour, red wine, oil, baking powder, anise seed, baking soda, and 1 cup of the sugar. Mix on medium speed until a dough forms. Scrape the dough out onto a clean work surface. Pour the remaining sugar on a plate.

3. To form the cookies, cut off a walnut-size piece from the dough and roll it out to make a rope 4 inches long. Form it into a ring and pinch the ends to close it. Dip one side of the cookie into the granulated sugar and then transfer to the prepared baking sheets, sugar side facing up.

4. Bake the cookies, one sheet at a time, until they're set and lightly golden on the bottom, 15 to 20 minutes. Transfer to a wire rack and let cool completely.

MAKES ABOUT 40 COOKIES

Red Wine Cookies

Jam-Filled Cookies

BISCOTTI RIPIENI DI MARMELLATA

Jam crostatas are a mainstay throughout all of Italy. Here the same flavors come together in a miniature version made into cookies.

½ cup (1 stick) unsalted butter, softened

⅓ cup superfine sugar

1 large egg

1¾ cups (225 g) all-purpose flour

Pinch of table salt

½ tablespoon grated lemon zest

1 teaspoon (5 ml) vanilla

½ cup fruit jam, such as grape or apricot

1. In a stand mixer with a paddle attachment, cream together the butter and sugar on medium speed until light and fluffy, about 3 minutes. Add the egg and mix until combined. Reduce the speed to low and mix in the flour, salt, lemon zest, and vanilla until a dough forms. Turn the dough out onto a work surface dusted lightly with flour and knead until smooth. Flatten the dough into a disk and wrap in plastic wrap. Refrigerate the dough for 1 hour.

2. Position a rack in the center of the oven and heat the oven to 350°F (180°C).

3. On a lightly floured work surface, roll the dough out into a large circle ⅛ inch thick. With a 3-inch round cookie cutter, preferably fluted, cut out disks of dough. You can gather the scraps and reroll the dough once to make more cookies. Put 1 teaspoon of jam in the center of each round, fold one side into the center, and bring the opposite side into the center. Arrange on a baking sheet lined with parchment paper, then refrigerate for 30 minutes.

4. Bake until the cookies are golden brown and set, 15 to 20 minutes. Transfer to a wire rack and let cool completely.

MAKES 22 COOKIES

Biscotti con Marmellata

2 cups (473 ml) heavy cream

1¼ cups (300 ml) whole milk

2 1-inch strips orange zest, white pith removed

3 large eggs

2 egg yolks

½ cup plus 3 tablespoons granulated sugar

¼ cup cornstarch

3 tablespoons unsalted butter, cut into pieces

4 cups small red wine grapes

1 recipe crepes (with milk) (see "Basic Recipes")

Cream-Filled Crepes with Roasted Grape Sauce

CRESPELLE CON CREMA E SUGO D'UVA

At Casale Verdeluna, Lino has strict criteria for selecting grapes designated for winemaking; those bunches that he passes over make their way into the kitchen for making jams and jellies. In this recipe, the grapes are roasted with sugar until they break down and become jam-like. Then they're spooned over the cream-filled crepes.

1. Add 1 cup of the cream, the milk, and the orange zest to a 3-quart saucepan over medium heat. Heat the mixture until it is just about to boil and then remove from the heat. In a large bowl, whisk together the eggs, yolks, ½ cup of the sugar, and the cornstarch until combined. Gradually whisk in the cream mixture. Pour the mixture back into the pan and set over medium-low heat. Cook, stirring constantly with a wooden spoon, until the mixture is thick enough to coat the back of the spoon, 4 to 8 minutes. Stir in the butter. Pass the cream through a fine-mesh sieve into a medium bowl and then place a piece of plastic wrap directly onto the surface. Refrigerate until completely chilled, 3 to 4 hours.

2. In a stand mixer fitted with a whisk attachment, beat the remaining cup of heavy cream at medium speed until medium-stiff peaks form, 3 to 5 minutes. With a spatula, fold the whipped cream into the pastry cream. Cover with plastic wrap and refrigerate for 30 minutes.

3. Position a rack in the center of the oven and heat the oven to 425°F (220°C).

4. In a large bowl, toss the grapes with the remaining 3 tablespoons of sugar and then transfer to a baking dish. Roast the grapes in the oven until they release their juices and begin to break down, 25 to 30 minutes. Transfer to a wire rack and let cool. Put the grapes and their juices into a food processor and pulse to make a chunky sauce.

5. To assemble the crepes, lay a crepe on a flat, clean work surface and spread ¼ cup of pastry cream over the bottom half. Fold the top half down to cover the cream and transfer to a dessert plate. Finish with the remaining crepes. Drizzle each crepe with the grape sauce and serve.

SERVES 12

Cesanese wine grapes

LE MOLE SUL FARFA

Tourists flock to the heart of Rome to see the splendor of the Roman dynasty's still-standing structures. But among the Eternal City's modern-day chaos, it is often difficult to view the architectural ruins and reflect upon life thousands of years ago. Beyond the city walls and past the ever-growing metropolitan sprawl, there exists an unspoiled, living testament to the Roman dynasty, in an area known as the Sabina. Here Stefano Fassone and his wife, Elizabeth, run their quaint olive farm and agriturismo Le Mole sul Farfa.

From the farmhouse, a dirt path called the strada Romana leads to an ancient grove of olive trees that are Stefano's pride and joy. Walking among them reveals their age and beauty, but hearing Stefano recant their fruit-bearing history is what really makes them special. Stefano has learned that some of his trees date back to Roman times, and a tour of his property provides compelling evidence. When he first purchased the olive grove from two shepherding brothers, Stefano began cleaning out the animal stall beneath the small house where he and Elizabeth currently live. The ground was covered in a thick layer of petrified sheep and goat manure. Excavating this revealed the remains of a Roman villa. Encouraged, Stefano continued to dig until he unearthed an ancient stone wheel and—to his amazement—an intact three-bay well. Apparently after olives were pressed, oil would flow into two of these bays, where the water would separate and rise to the surface. It was then skimmed, and the oil was transferred to the third bay, where sedi-

ment could fall to the bottom and oil could be transferred to urns.

This archaeological discovery led Stefano out to take a closer look at the individual trees around the property. He knew that their gnarled knotty trunks, distorted branches, and pockmarked trunks all indicated their age, but now he believed them to date back to Roman varietals.

To complete his modern-day living olive grove museum, Stefano introduced to his farm a breed of donkeys whose lineage dates back to 1000 BC. They graze lazily beneath the trees with sheep and goats, feeding on the weeds and grass and fertilizing the soil. All of this creates a natural ecosystem that parallels the farming methods in practice thousands of years ago. Today, strolling among the ancient trees and the quietly feeding animals while looking out at the untouched natural beauty of Sabina's rolling hills lends a majestic feeling of what life must have once been like on a Roman estate.

The love Stefano and Elizabeth have for their land, and their commitment to sustainability led them to petition for it to become a regional park. Today their corner of the Sabina is documented as a protected area and an outdoor lover's destination. Stefano and Elizabeth maintain a network of hiking trails that wander through the park. A favorite trek from the farm leads down to the crystal-clear waters of the sparkling Farfa River. The solar panels that provide heat and hot water to the agriturismo and the vegetarian restaurant of Le Mole sul Farfa reflect the ecofriendly approach Stefano and Elizabeth have adopted.

Originally from Belgium, Elizabeth spent time learning

local recipes in the kitchens of neighboring women, who taught her the specialties of the Sabina. Today she cooks the way Italians did when meat was scarce and eating was based on what the land could provide. In her all-vegetarian kitchen, she prepares both traditional recipes taught to her by her neighbors as well as more modern dishes she has learned on her own. Protein-rich chickpea soup with irregular-shaped pasta; a non-Italian risotto made from ginger, lemon, and squash; nutrient-rich buckwheat crepes filled with onions, feta, walnuts, and arugula; and puff pastry pockets stuffed with grilled radicchio, pear, and creamy Taleggio all make for a matchless dining experience. Bottles of verdant green extra-virgin olive oil line the table—Stefano's single-varietal oils from particular trees. Small amounts are poured into paper cups for guests to sample while Stefano inquires into their flavors and aromas, much as at a wine tasting. When dinner arrives, bottles are left for drizzling over each dish. The strong, slightly bitter and spicy notes of these olive oils may be the last living legacy of the Roman Empire.

Roman ruins of olive cistern

Cheese Biscuit Napoleons with Green Apples

STUZZICHINI AL FORMAGGIO E MELE VERDI

These are great party appetizers. The creamy combination of three cheeses combines with a sour tinge from the green apple to complement the flaky texture of the biscuit.

For the dough:

2/3 cup (100 g) all-purpose flour, plus more for rolling

1 3/4 ounces (50 g) ricotta cheese

1/4 cup (1/2 stick) unsalted butter, softened

1 tablespoon grated Parmigiano Reggiano

Kosher salt

For the filling:

2 ounces (1/4 cup/60 g) mascarpone

2 ounces (2 1/2 tablespoons/60 g) ricotta cheese

1 ounce (30 g) goat cheese

1 teaspoon (5 ml) extra-virgin olive oil

1 teaspoon chopped chives, plus more for garnish

Pinch of cayenne pepper

Kosher salt and freshly ground black pepper

1. **Make the dough:** In a stand mixer fitted with a paddle attachment, add the flour, ricotta, butter, Parmigiano, and 1/2 teaspoon of salt. Mix on medium speed until a dough forms. Wrap the dough in plastic wrap and flatten into a square. Refrigerate the dough for 1 hour.

2. **Make the filling:** In a medium bowl, mash the mascarpone, ricotta, and goat cheese together with a fork until smooth. Mix in the oil, chives, and cayenne; season to taste with salt and pepper. Cover with plastic wrap and refrigerate until needed.

3. **To assemble:** Position a rack in the center of the oven and heat the oven to 350°F (180°C). Line a baking sheet with parchment paper.

4. Lightly dust a work surface with flour and roll out the dough into a 10-inch square 1/8-inch thick. With a fluted pastry cutter, cut the dough into 20 squares. Transfer to the prepared baking sheet.

5. In a small bowl, beat together the egg yolk with 1 teaspoon of water. Brush the egg wash over the biscuits and top 10 of them with a walnut half. Bake in the oven until golden brown and puffed, 8 to 12 minutes. Remove from the oven and let cool on a wire rack for 5 minutes.

6. Place two biscuits without walnut pieces on an appetizer plate. Spread 1 tablespoon of cheese filling over each biscuit, and then sprinkle with diced apple. Top each with a biscuit containing a walnut, with the walnut on the outside. Garnish with some chives. Continue assembling the remaining biscuits.

SERVES 5

To assemble:

1 egg yolk

10 walnut halves

1 Granny Smith apple, cored and
  cut into fine dice

Grazing time

Puff Pastry Pockets Stuffed with Radicchio and Pear

FAGOTTINI DI RADICCHIO E PERE

The sweetness of the pear helps cut the bitterness of the radicchio in this dish, while the Taleggio adds a hint of nuttiness. These pockets make great appetizers; or you can serve them with a side salad for a light lunch or first course.

2 small heads radicchio Trevisano, quartered lengthwise with core intact

1 small or ½ medium Bosc pear, peeled, cored, and quartered lengthwise

2 tablespoons (30 ml) extra-virgin olive oil

Kosher salt and freshly ground white pepper

1 medium shallot, finely chopped

2 small cloves garlic, lightly smashed

2 tablespoons (30 ml) white wine

2 teaspoons (10 ml) red wine vinegar

All-purpose flour, as needed

1 pound (453 g) puff pastry dough

1 egg yolk

2¼ ounces (75 g) Taleggio cheese, rind removed and cut into ¼-inch pieces

6 tablespoons grated pecorino cheese

1. Heat a grill pan or a heavy cast-iron skillet over medium-high heat. Brush the radicchio and pear quarters all over with 1 tablespoon of the olive oil and season with salt and pepper. Grill the radicchio quarters until lightly browned, 2 to 3 minutes. Flip them over and grill until browned and just tender, 2 to 3 minutes more. Transfer to a cutting board. Grill the pear quarters until golden brown, about 3 minutes. Turn the pears over and continue grilling until browned and tender, 3 to 4 minutes. Transfer to the cutting board.

2. Remove the core from each radicchio quarter and then chop up the radicchio. Cut the pear quarters into ¼-inch dice and transfer to a bowl.

3. Heat the remaining 1 tablespoon of oil in a 12-inch skillet over medium heat. Add the shallot and garlic cloves and cook until golden and fragrant, 1 to 2 minutes. Discard the garlic cloves. Add the radicchio to the pan and cook, stirring occasionally, for 3 minutes. Add the white wine and red wine vinegar and cook until the liquid has reduced to dry, 6 to 8 minutes. Season to taste with salt and pepper. Spread the mixture out on a plate and let cool to room temperature.

4. Position a rack in the center of the oven and heat the oven to 400°F (200°C). Line a baking sheet with parchment paper.

5. On a lightly floured surface, roll out the puff pastry to a 12 × 14-inch rectangle. Cut the dough into 3 × 3½-inch rectangles.

6. In a small bowl, beat together the egg yolk with 1 teaspoon of water. Brush the edges of each puff pastry rectangle with some of the egg wash.

7. Divide the radicchio mixture among the rectangles, and then top each with a few pieces of pear, some Taleggio, and a heaping teaspoon of pecorino. Roll the rectangles up lengthwise and seal the edges well. Transfer to the prepared sheet tray, prick the tops with the tines of a fork, and then brush with the remaining egg wash. Bake in the oven until deep golden brown and the dough puffs, 20 to 24 minutes. Let cool slightly before serving.

MAKES 16 POCKETS

Puff Pastry Pockets Stuffed with Radicchio and Pear

Buckwheat Crepes with Feta, Red Onions, and Arugula

CRESPELLE DI GRANO SARACENO CON FORMAGGIO DI CAPRA, CIPOLLA ROSSA, E RUCOLA

This is a staple on Elizabeth's vegetarian menu that she makes year-round. The stuffed crepes are healthy and balanced with nutrients and her buckwheat flour batter adds texture and a hint of nutty flavor.

For the filling:

2 tablespoons (30 ml) extra-virgin olive oil

6 tablespoons (¾ stick) unsalted butter

3 medium red onions, halved and thinly sliced

Kosher salt

1 teaspoon chopped fresh thyme

5 cups baby arugula

¾ cup walnuts, finely chopped

1 medium shallot, finely chopped

¼ cup (59 ml) balsamic vinegar

Freshly ground black pepper

For the crepes:

½ cup (65 g) buckwheat flour

½ cup (62.5 g) all-purpose flour

1 large egg plus 1 yolk

1½ cups (360 ml) whole milk

1 tablespoon (15 ml) extra-virgin olive oil, plus more for the pan

Pinch of kosher salt

To assemble:

4 ounces (about 1 cup) crumbled feta

1. **Make the filling:** Heat the olive oil with 1 tablespoon of the butter in a 12-inch skillet over medium heat. Add the onions and a generous pinch of salt and cook, stirring occasionally, until the onions are very tender and beginning to brown, 15 to 20 minutes. Remove from the heat. Stir in the thyme and arugula and toss until the arugula wilts.

   Put the remaining butter and the walnuts in a 2-quart saucepan over medium heat. Cook until the butter begins to brown and the walnuts are toasted, 3 to 5 minutes. Stir in the shallot, balsamic vinegar, and a few grinds of black pepper, and cook until the flavors meld, 1 to 2 minutes. Add 2 tablespoons of the sauce to the onion mixture and toss well to combine. Keep the remaining sauce warm.

2. **Make the crepes:** In a large bowl, mix together the buckwheat and all-purpose flours. Whisk in the egg, egg yolk, milk, 1 tablespoon of oil, and salt until a smooth batter forms. Heat a 10-inch nonstick skillet over medium-high heat, brushing it with a little oil. Pour ¼ cup of the batter into the skillet. Swirl so the batter thinly and evenly coats the base of the pan. Cook until the crêpe is spotted with brown on the underside, about 1 minute, then flip and cook the other side until lightly browned, 30 seconds to 1 minute more. Transfer the crepe to a plate and continue cooking the crepes in this manner, oiling the pan as necessary.

3. **To assemble,** fill each crepe with 1 tablespoon crumbled feta and ¼ cup of the onion mixture. Fold in half and then in half again. (They should have a triangular shape.) Transfer to individual plates and drizzle with the remaining sauce.

SERVES 6

MONO-VARIETAL OLIVE OIL

In 1954 a devastating winter storm swept throughout central Italy. Snow and ice piled up while the cold penetrated deep into the earth, splitting vines and tree trunks, destroying vineyards and crippling olive groves. Nearly every region was hit by the storm, and in the spring farmers were forced to replant their land with new saplings. Many chose to narrow the selection of olive trees and preferred types known for bearing large quantities of fruit, eliminating ancient strains indigenous to the area.

The new generation of olive oil that resulted was soft and smooth and lacked the bitter and peppery bite of years past. Over time this neutral flavor became accepted as the standard for high-quality extra-virgin olive oil. However, a few olive groves were spared this fate, including Le Mole sul Farfa, whose deep-rooted trees survived the storm. Today Stefano individually harvests each tree to produce more than ten mono-varietal oils, each displaying its own individual flavor and personality. The oils are complex and unique and range from super-spicy, to astringently bitter, to light and smooth. The farm also produces a blend from all of the different olives, and Stefano can adjust the ratios he uses depending upon the flavor profile he's looking for. All of the grove's oils are presented in Le Mole sul Farfa's restaurant with various dishes. Their true flavors outshine the bland modern-day oils found on supermarket shelves.

Fettuccine with Almond Pesto and Cherry Tomatoes

STRINGOZZI AL PESTO DI MANDORLE E CILIEGINO

½ cup blanched almonds

15 cherry tomatoes, quartered

1 clove garlic, finely chopped

½ cup grated pecorino cheese

Kosher salt

12 ounces (340 g) fettuccine or homemade stringossi pasta

¼ cup lightly packed basil leaves

Extra-virgin olive oil, for serving

Making stringossi—a type of pasta from the Sabina—involves an intricate shaping technique that is difficult to master, especially for an outsider not born into a family of experienced pasta makers. Elizabeth wanted to learn the secrets involved in making the pasta, so she sought the help of a townswoman from nearby Mompeo who'd lived in the area for fifty years. When the other local women heard that someone not born and bred in the village had taught Elizabeth, they insisted that she hadn't been taught the true recipe. Elizabeth laughs when she tells this story: If someone who has lived in the area for over five decades is still viewed as an outsider . . . she will always be an outsider.

1. In a food processor fitted with the blade attachment, pulse the almonds until finely chopped. Transfer the almonds to a large bowl and mix in the cherry tomatoes, garlic, pecorino, and a pinch of salt.

2. Bring a large pot of well-salted water to a boil. Drop in the pasta and cook until al dente, following the cooking instructions on the package. Reserve ½ cup of water and then drain the pasta well. Return the pasta to the pot, adding the pesto and ¼ cup of the reserved pasta water. Toss well to combine, adding more water 1 tablespoon at a time if the sauce seems dry. Toss in the basil leaves and then divide among four large shallow bowls. Drizzle each serving with a little oil and serve.

SERVES 4

Walnut Honey Brittle
NOCIATA

Bay leaves typically flavor all types of savory dishes throughout Italy. Here, Elizabeth showed us the herb's versatility by teaching us this traditional bay dessert of the Sabinia. The key to eating it is to peel back the bay leaves slightly, allowing their aroma to mingle with the sweetness and tannins of the walnut brittle. You can wash the leaves afterward and reuse them for cooking—or, as they do at Mole sul Farfa, you can toss them into the fire for a mini firework display of the oils and sugars igniting.

2/3 cup honey

8 ounces (226 g) finely chopped
    walnuts

100 fresh bay leaves

1. Put the honey in a 3-quart pan over medium heat and cook until it's very liquidy, 1 to 2 minutes. Stir in the walnuts, and cook until the walnuts are toasted, the honey begins to caramelize, and the mixture pulls away from the sides of the pan, 8 to 10 minutes.

2. Lightly dampen a baking sheet with water and pour the walnut mixture onto it. Dampen an offset spatula with water and spread the mixture out. With a dampened meat mallet, lightly push down on the mixture to flatten it evenly into a 13½ × 9½-inch rectangle.

3. With a sharp knife, cut the mixture into 2-inch strips lengthwise, and then cut it into 2-inch strips crosswise on an angle to create diamonds. Using the offset spatula, lift up one of the diamonds and put it on the matte side of a bay leaf so the points of the diamond are facing the points of the leaf. Top with the matte side of another bay leaf and press the leaves together to adhere to the walnut brittle. Transfer to a large plate. Continue with the remaining pieces. Refrigerate for at least 8 hours and up to overnight.

MAKES 50 PIECES

Walnut Honey Brittle

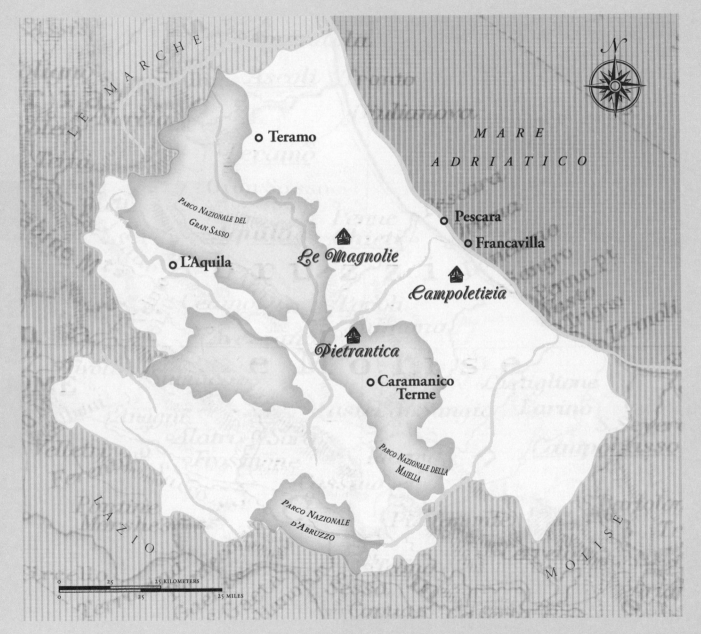

ABRUZZO

LE MARCHE

LAZIO

MOLISE

MARE ADRIATICO

o Teramo

Parco Nazionale del Gran Sasso

o L'Aquila

Le Magnolie

o Pescara
o Francavilla

Campoletizia

Pietrantica

o Caramanico Terme

Parco Nazionale della Maiella

Parco Nazionale d'Abruzzo

N

0   25   25 KILOMETERS
0   25   25 MILES

CHAPTER 6

ABRUZZO

LE MAGNOLIE

For a brief period each autumn, Le Magnolie's garden pops, with an explosion of small violet crocuses. Packed into tidy, organized rows, the flowers bloom in the early morning, exposing the fire-engine-red stigmas of saffron, one of Abruzzo's most ancient and coveted crops. We happened to arrive at Le Magnolie during their harvest, and on a warm, sun-kissed early-autumn morning, we joined owner Gabrielle to lend a hand. With knees popping and bodies contorted in hunched positions, we learned the finer, backbreaking intricacies of growing this delicate crop, as each individual flower must be individually plucked and placed gently in a basket to avoid crushing the precious spice. Once collected, the meticulous process of separating the stigmas from the blossoms began, staining fingers a burnt orange hue. The stigmas are then dried in a low oven, and their aromatic scent permeates the farmhouse, greeting guests with a rich, warming aroma.

Husband Mario and wife Gabrielle Tortella are pioneers in the Abruzzo, recognized as some of the first in the region to implement all-organic growing methods for their farm. The couple is also credited with establishing their farm as one of the first agriturismi to open in the area, and it has enjoyed great success throughout its twenty-year tenure. Le Magnolie has established a loyal clientele of both Italian and international guests, as well as deep roots within the community. Gabrielle's position as a chairman with the agricultural board of Abruzzo connects her to other small family farmers throughout the region, and she works with other artisanal-minded winemakers, dairy farmers, truffle hunters, butchers, and pasta makers to source her ingredients.

In a dining room with a vaulted brick ceiling and walls adorned with hand-painted ceramics, guests sit together at one long narrow table. The atmosphere is warm and inviting and reflective of Gabriella and Mario,

Olga cooking

who offer glasses of wine to their guests and encourage folks to wander into the kitchen to see what Gabriella and her mother Olga are preparing. The farm's position in the rolling foothills of the Gran Sasso, Abruzzo's tallest peak, affords a long growing season suited for a variety of vegetables including tomatoes, beans, fennel, artichokes, cardoons, peppers, eggplant, swiss chard, and zucchini. This bounty forms the basis for plates of shared antipasti that are passed around the table: warm focaccia topped with peppers and eggs, stewed swiss chard with borlotti beans, pan-fried cheese topped with slow-roasted tomatoes, and thick wedges of pecorino drizzled with Le Magnolie's honey and chopped walnuts.

With one of the world's most precious ingredients growing out the kitchen door, Gabrielle infuses saffron into both savory and sweet dishes, matching it with locally grown farro as well as infusing it into a cream base for a delectable panna cotta. Long-simmered sauces made from Le Magnolie's own ducks and geese are particularly delicious when paired with the toothsome chitarra pasta, made from thick sheets of pasta rolled over wire guitar strings. Layers of crepes are covered with a pork-rib ragù and baked in the oven for an Abruzzese lasagna known as a timballo. Small chunks of mutton and fat are threaded onto wooden skewers and grilled over a wood fire for a second course that highlights the quality of meat from sheep raised in the region's verdant mountain pastures. Mario's cherished olive trees provide neon green liquid gold that appears in every dish. In addition to drizzling over salads and sautéing, olive oil is used for frying as well

as for desserts, in which it replaces butter for a lighter and somewhat healthier preparation. For each course, guests help themselves from platters, which sets the tone for a convivial evening amongst strangers. A quaffable yet serious Montepulciano d' Abruzzo red wine from a small award-winning vineyard down the street fuels worldly debates long into the night. A look into Le Magnolie's guestbook reveals the genuine affection guests have for the farm and for Mario and Gabriella. No doubt this list of returning couples and families will continue to grow year after year and well into the farm's prosperous future.

Harvesting saffron

Campo Imperatore

The Gran Sasso National Park holds a treasure of small preserved medieval towns and robust wilderness. A windy road from the valley below leads through dense forests and past sleepy villages, until it reaches the enormous expanse of the alpine plateau known as Campo Imperatore. The landscape opens to grass-covered plains and ridges; the cloud-covered peak of the Gran Sasso, central Italy's highest mountain, looms in the distance. This virgin landscape and unique topography has been called Italy's "little Tibet" and has been the backdrop for many movies, including several of America's most famous westerns. Throughout summer and into early autumn, the fields are speckled with cows and sheep, freely grazing on the pastures and adding a touch of pastoral beauty to the dramatic setting. Building is prohibited in the park, except for a small ski resort and two wooden shacks that cook up simple plates of grilled meats. On offer are various cuts of pork and steak—but arrosticini, skewers of succulent lamb, are the true Abruzzese specialty. Summer days see both restaurants packed, and a festive scene prevails. Diners sit outside at long picnic tables next to open-fire grills, grazing animals, and a never-ending panorama of stunning beauty.

Creamy Farro with Saffron and Toasted Pine Nuts

FARRO ALLO ZAFFERANO CON PIGNOLI TOSTATI

2½ quarts (2.5 liters) homemade chicken broth or canned low-sodium broth

½ teaspoon saffron

2 tablespoons (30 ml) extra-virgin olive oil

1 small yellow onion, cut into fine dice

Kosher salt

2 cups (13 oz./380 g) pearled farro

½ cup (118 ml) dry white wine

2 tablespoons (¼ stick) unsalted butter

¼ cup grated Parmigiano Reggiano, plus ¼ cup shaved for garnish

Freshly ground black pepper

⅓ cup pine nuts, toasted

Mario and Gabrielle harvest their own saffron each fall at Le Magnolie. The process requires patience, flexibility, and a strong back, as picking the small crocuses involves hours spent hunched over the long rows of flowers. Here, Gabrielle pairs the spice with nutty farro and pine nuts to complement its delicate nature.

1. Bring the chicken broth to a simmer in a 4-quart pot over medium heat. Reduce the heat to low and keep warm. Combine ½ cup of warm broth with the saffron in a small liquid measuring cup and set aside to let the saffron bloom.

2. In a 12-inch straight-sided sauté pan, heat the olive oil over medium-high heat. Add the onion and a pinch of salt, and cook, stirring occasionally, until tender and just beginning to brown, 5 to 6 minutes. Add the farro to the pan and stir until the grains are coated with the oil and lightly toasted, 1 to 2 minutes. Add the wine and cook, stirring occasionally, until reduced to dry, about 30 seconds.

3. Add the saffron broth to the pan and ladle enough of the warm broth into the pan to barely cover the farro. Bring to a boil, then adjust the heat to maintain a lively simmer. Cook, stirring frequently, until the broth is mostly absorbed, 2 to 3 minutes. Continue adding broth a ladleful at a time, stirring occasionally and letting each addition be absorbed before adding the next. Cook until the farro is toothsome and creamy, 20 to 25 minutes. Stir in the butter and ¼ cup of the grated Parmigiano; season to taste with salt and pepper.

4. Serve the farro in individual shallow bowls and top with the toasted pine nuts and shavings of Parmigiano.

SERVES 4

Abruzzese Crepe Lasagna with Pork Ragù

TIMBALLO ABBRUZZESE

3½ tablespoons (51 ml) extra-virgin olive oil

2 pounds (900 g/3–4 ribs) bone-in country style pork ribs

Kosher salt and freshly ground black pepper

1 small yellow onion, cut into fine dice

1 carrot, peeled and cut into fine dice

1 stalk celery, cut into fine dice

1 2-inch sprig rosemary

1 bay leaf, preferably fresh

2 28-ounce (793 g) cans whole plum tomatoes, finely chopped with their juices

1 recipe crepes (see "Basic Recipes")

½ cup grated Parmigiano Reggiano

12 ounces (340 g) fresh mozzarella, shredded

This is Abruzzo's signature lasagna, which uses crepes instead of pasta. At La Magnolie rows of crepes are layered with mozzarella and a slow-cooked tomato and pork ragù, which is baked and served piping hot. The outer edges of the timballo become dark and crispy while the center crepes are pleasantly soft and saturated with flavorful sauce.

1. Heat 2½ tablespoons of the olive oil in a large saucepan over medium-high heat. Season the pork all over with salt and pepper; add to the pan and brown evenly on all sides, 3 to 4 minutes per side. Remove from the pan and set aside on a plate. Pour off all but a thin layer of fat from the skillet and add the remaining 1 tablespoon olive oil. Add the onion, carrot, celery, and a pinch of salt and cook until the vegetables are very tender and browned, stirring occasionally, 3 to 4 minutes. Add the rosemary and bay leaf and cook until fragrant, about 30 seconds.

2. Return the ribs to the pan. Add the tomatoes and ½ cup of water. Bring to a boil, reduce the heat to maintain a very gentle simmer, and cook until the meat is fork-tender, 2½ to 3 hours.

3. Remove from the heat and let the sauce cool. Remove the ribs from the sauce. Using two forks, shred the meat from the bones and put into a medium bowl. Discard the bones.

4. Position a rack in the center of the oven and heat the oven to 425°F (220°C).

5. Assemble the lasagna in a 9 × 13 × 3-inch baking dish. Spread ½ cup of the sauce in a thin layer on the bottom of the dish. Then cover with a slightly overlapping layer of crepes, cutting them as needed to fill the gaps. Spread ½ cup of the sauce over the crepes. Sprinkle with 1 tablespoon Parmigiano, ½ cup shredded pork, and ½ cup mozzarella. Add a new layer of crepes, overlapping them slightly.

6. Repeat the layers as instructed above, to make a total of four layers. For the final layer, spread the remaining ½ cup of sauce over the top of the crepes, and then sprinkle with 2 tablespoons of Parmigiano.

7. Bake until heated through and bubbling at the edges, 30 to 40 minutes. Remove from the oven and let rest for 10 to 15 minutes before serving.

SERVES 8–10

Oven-Roasted Tomatoes

POMODORI AL FORNO

Try this recipe during peak tomato season for a tasty summer side dish. The tomatoes can also be left to cool and added to salads, used as a pasta sauce, or used as a filling for sandwiches.

¼ cup finely chopped parsley

2 small cloves garlic, minced

1 pound (453 g) small tomatoes, cut in half crosswise

Kosher salt and pepper

2 tablespoons (30 ml) extra-virgin olive oil

SERVES 6

1. Position a rack in the center of the oven and heat the oven to 375°F (190°C).

2. In a small bowl, combine the parsley and the garlic. Arrange the tomato halves in a 9 × 13-inch baking dish. Season with salt and pepper and drizzle with the oil. Roast in the oven until the tomatoes are just tender, about 12 minutes.

3. Remove the tomatoes from the oven and sprinkle evenly with the parsley mixture. Baste with any juices from the pan, and return to the oven. Roast until the tomatoes are very tender, 20 to 25 minutes. Turn the broiler to high and cook the tomatoes until browned and blistery, 4 to 6 minutes.

Oven-Roasted Tomatoes

Saffron-Infused Panna Cotta
PANNA COTTA ALLO ZAFFERANO

The high fat content in heavy cream helps bring out the aromatic flavor of saffron, and its use as a main ingredient in a dessert shows the versatility of the spice. At Le Magnolie this creamy treat is a staple during the saffron harvest.

1 cup (240 ml) milk
¼ teaspoon crumbled saffron threads
2 teaspoons powdered gelatin
½ cup granulated sugar
Pinch of table salt
3 cups (720 ml) heavy cream
½ teaspoon vanilla extract
2½ tablespoons (2.36 ml) mild honey
¼ cup chopped toasted walnuts

1. Pour the milk into a 2-quart saucepan. Add the saffron and then sprinkle the gelatin evenly over the surface. Let stand for 10 minutes. Warm the milk over low heat, add the sugar and salt, and stir until the sugar is completely dissolved, 2 to 3 minutes.

2. In a large bowl, combine the heavy cream and vanilla. Whisk in the warm milk until combined. Divide the mixture among eight 4-ounce ramekins, cover with plastic wrap, and refrigerate until set, at least 4 hours.

3. To serve, unmold the panna cotta by running a butter knife between the custard and the ramekin. Invert each panna cotta onto an individual plate and carefully remove from the ramekin. Drizzle the top of each with 1 teaspoon honey and sprinkle with the walnuts.

SERVES 8

Chocolate Almond Cream Cakes

BOCCONOTTI

Gabrielle uses extra-virgin olive oil when making her bocconotti, yielding a tender, flaky crust. Once they're baked, she dusts them heavily with confectioner's sugar, laughing with guests over their powdered covered lips and shirts.

For the dough:

5 egg yolks

5 tablespoons granulated sugar

5 tablespoons (73 ml) extra-virgin olive oil

1 tablespoon grated lemon zest

1¾ cups (225 g) all-purpose flour, plus extra for the molds

For the filling:

1 cup (240 ml) water

3 tablespoons granulated sugar

2 ounces (60 g) dark chocolate, chopped

1¾ ounces (49 g) toasted almonds, coarsely ground

2 egg yolks

⅛ teaspoon ground cinnamon

To assemble:

Extra-virgin olive oil, as needed

Powdered sugar, as needed

1. **Make the dough:** In a stand mixer fitted with the paddle attachment, combine the egg yolks and sugar. Mix on medium speed until thickened and pale yellow, 4 to 5 minutes. Add the oil and lemon zest and mix until combined. Add the flour, and mix on low speed until a dough starts to form. Turn the dough out onto a lightly floured surface and knead until smooth, 1 to 2 minutes. Put the dough back into the bowl of the stand mixer, cover with plastic wrap, and set aside at room temperature.

2. **Make the filling:** Combine the water and sugar in a 4-quart pot over medium heat. Bring to a boil, stirring often, and cook until the sugar dissolves. Add the chocolate and continue cooking, stirring continuously until the chocolate is melted, about 1 minute. Remove from the heat and stir in the ground almonds. Let the mixture cool slightly. Whisk in the egg yolks, one at a time, until combined. Put the pan over medium-low heat, and cook, stirring continuously, until the mixture has thickened, 3 to 5 minutes. Stir in the cinnamon, remove from the heat, and let cool completely.

3. **To assemble,** position a rack in the center of the oven and heat the oven to 350°F (180°C).

4. Brush ten 2-inch tart molds with oil and lightly dust with flour. Cut off a tablespoon of dough and press into a tart pan, coming slightly over the edge. Fill the tart pan with a heaping tablespoon of the chocolate filling.

Cut off another tablespoon of dough and gently flatten it with the palm of your hand to a 3-inch round. Cover the filling with the dough round and then seal the edges of the dough with your fingers. Trim away any excess dough. Continue with the remaining molds.

5. Arrange the tart pans onto a rimmed baking sheet and bake in the oven until the dough is lightly golden brown and set, 25 to 30 minutes. Remove from the oven and let cool for 5 minutes on a rack. Remove the tarts from the tart pans and cool completely on a rack. To serve, generously dust each tart with powdered sugar.

MAKES 10 INDIVIDUAL CAKES

Chocolate Almond Cream Cakes

CAMPOLETIZIA

Campoletizia's farm heritage began shortly after the Second World War, when an engineer from the Veneto purchased the property as an investment. After his passing, the property was divided between his children, and his daughter, Letizia, chose to continue farming the portion of land she inherited, changing its main production from olive oil and wine to raising pigs for porchetta and opening a butchery. Her son Nicola came to work there, leaving his studies in medicine to pursue his true passion, agriculture. He quickly developed a vision and plan for changing the direction of the farm, and they sold their successful pork business to open an agriturismo. Today Letizia, Nicola, and his wife, Livia, have opened a unique and hip country retreat that has become a favorite destination for Romans seeking a weekend escape from city life, locals who stop by on weekends for leisurely meals, and foreigners who have become enchanted with the pleasures of the working farm.

No detail has been spared in the design and furnishing of the agriturismo, whose original style marries rusticity with urban flair. Country antiques with chipped paint and worn wood fit perfectly into the modern style and clean lines of the restaurant and guest rooms. From little ball jars that sit on the table filled with wildflowers, to stylish plates and atypically shaped glasses, to the presentation of food in terra-cotta dishes representative of the Abruzzo, attention to fine details creates Campoletizia's immense appeal. In the dining room, still lifes of food and bucolic country scenes painted by Livia's ninety-five-year-old grandmother adorn the walls, an old pharmacist's chest of drawers holds all the flatware, and a stack of old wooden apple crates serves as shelving for all of the farm's products for sale, while a polished gray concrete floor, white-beamed ceiling, and red-trimmed windows complete the space.

But behind the stylish interior design, the real key to Campoletizia's success lies within the dedication and passion Nicola and Livia bring to their farm. Part of a younger generation of Italians focused on preserving tradition, the couple has breathed new life and energy into a nearly extinct lifestyle. The farm itself has become streamlined and has evolved a sustainable production that follows the seasons to generate an assortment of goods for use in the restaurant. Olive oil pressed at an old-fashioned mill, wine, free-range native black pigs, chickens, organic fruits and vegetables, and revived antique grains all make their way onto the table in various preparations. The pair's good natures and happiness are obvious everywhere at Campoletizia, heightened by gently sloping hills aflame with color that encircle the property.

Letizia and Livia cook together in a small kitchen that runs nonstop on weekends, when the restaurant opens its doors to guests not staying at the farm. Its popularity has surged since opening, and tables are hard to come by, especially on Sundays, when entire families descend on the farm to spend the afternoon. The definitive Italian Sunday lunch has tables piled high with platters of charcuterie, soft billowy fried dough with the essence of wild fennel or spicy with Abruzzese chile peppers, plates of aged sheep's cheese and jellies, and carafes of wine. Livia pops out of the kitchen clad in a retro printed apron carrying a terra-cotta bowl of the family's award-winning chicken lasagna, while Nicola carves pork roasts tableside and spoons on a succulent sauce made from their own grapes. The air is filled with the chaos and clamor of friends and family fueled by good food and wine. Weeknights find a more tranquil environment, and the restaurant is open to only overnight guests, who sit together and share meals with Nicola and Livia. This true essence of the Italian agriturismo is what Nicola enjoys most about Campoletizia, and as each course is presented, he describes the ingredients and their preparations, giving a sense of place and history to each dish. The array of classic family recipes Letizia has collected through the years and Livia's inventive twists are all nourishing and tasty, and the quality of their ingredients is evident in every bite. All are served family-style, in platters set on handcrafted metal stands. If you can pull your attention away from the delicious food and charisma of Nicola and Livia to take in the surroundings, the unique charm of this place becomes apparent.

Montepulciano d' Abruzzo grapes

Walnut-Stuffed Prunes Wrapped in Crispy Pancetta

PRUGNE SECCHE CON NOCI E PANCETTA

16 prunes, pitted

8 walnuts, cut in half

8 slices straight pancetta or bacon, cut in half

This is a great appetizer to serve in the cold winter months or at your next holiday feast. The saltiness of the pancetta pairs well with the sweetness of the prunes, and the walnut adds a nice tannic touch. You'll never look at prunes the same way after indulging in these.

1. Position a rack in the center of the oven and heat the oven to 400°F (200°C).

2. Make a slit in each prune large enough to fit a walnut piece. Stuff each with a walnut and then wrap with a slice of pancetta or bacon. Secure the pancetta with a toothpick. Arrange the prunes in a small baking dish and bake in the oven until the pancetta is crisp, 12 to 15 minutes. Remove from the oven and transfer to a serving platter.

MAKES 16

Shepherd-Style Fried Cheese Fritters
PALLOTTE CACI'E OVA

Taking their name from the Abruzzese dialect term for "cheese and egg," these crispy, creamy fritters are best served hot. The key here is the quality of the cheese: Opt for an aged pecorino to pay homage to the Abruzzo's shepherding heritage.

1. Cut the crust off the bread and discard or reserve for another use. Cut the bread into pieces, put into the bowl of a food processor, and pulse until coarse. Add the eggs, provolone, pecorino, Parmigiano, and baking soda; puree until smooth. Transfer to a large bowl, cover with plastic wrap, and refrigerate until completely chilled, about 1 hour.

2. Scoop up a generous teaspoonful of the mixture and shape it into a 1-inch ball with your hands. (If the mixture is sticky, lightly dampen your hands with a little olive oil.) Transfer to a rimmed baking sheet lined with parchment paper and continue shaping the balls.

3. Position a rack in the center of the oven and heat the oven to 200°F (90°C).

4. Heat ½ inch of canola oil in a large straight-sided skillet or cast-iron pan until shimmering hot. Add the fritters in batches and cook, turning occasionally, until crisp and brown on all sides, 4 to 6 minutes total. Transfer to an ovenproof plate lined with paper towels and keep warm in the oven while frying the remaining fritters. Serve immediately.

MAKES 24 FRITTERS

2½ ounces (75 g) rustic bread

3 large eggs

7 ounces (200 g) grated mild provolone

3½ ounces (100 g) grated aged pecorino

1¾ ounces (50 g) grated Parmigiano Reggiano

Pinch of baking soda

Canola oil, for frying

Shepherd-Style Fried Cheese Fritters

Chicken Lasagna from the Michetti Convent

LASAGNE DEL CONVENTO MICHETTI

1 whole chicken (about 3½ pounds/
    1½ kg)

Kosher salt and freshly ground white
    pepper

2 tablespoons (30 ml) extra-virgin
    olive oil

2 cloves garlic, smashed

¾ cup (175 ml) white wine

½ cup crushed tomatoes

1 small yellow onion, halved

1 small carrot, peeled and cut into
    thirds

1 stalk celery, cut into thirds

1 recipe egg pasta dough (see
    "Basic Recipes")

1½ cups grated Parmigiano
    Reggiano

An award-winning family heirloom recipe, this multi-layered chicken lasagna gets bathed in a fortified stock with a flavor reminiscent of chicken soup. It's a taste of comfort in every bite. A relative of the family first made the lasagna while working at an artists' convent in nearby Francavilla al Mare, and since then, the recipe has received accolades for its originality and absolute deliciousness. It took some serious begging and pleading on our part to be the first people outside Letizia's family to learn the secrets behind this heartwarming dish.

1. Using poultry shears, remove the backbone from the chicken by cutting along both sides. Extend the wings on each side and chop off the last two joints. Chop the backbone into three pieces and set aside with the wing tips. Cut the chicken into eight pieces and season all over with salt and pepper.

2. Heat the oil in a 12-inch heavy-duty skillet over medium-high heat. Add the chicken, skin-side down, and cook to a deep golden brown, about 4 minutes. Flip the chicken over, add the garlic to the pan, season with salt and pepper, and continue cooking until browned, 3 to 4 minutes. Add the wine to the pan and simmer until reduced by almost half. Stir in the crushed tomatoes and 3 cups of water. Bring to a boil, cover, reduce the heat, and simmer until the chicken is falling off the bone, 35 to 40 minutes. With a slotted spoon, transfer the chicken to a large plate and let cool. Strain the cooking liquid through a fine-mesh sieve into a large bowl. Let cool completely, then skim away any fat that rises to the surface.

3. When it's cool enough to handle, separate the meat from the skin and the bones. Shred the meat and set aside in a bowl. Put the skin and bones in a 4-quart saucepan. Add the backbone, wing tips, onion, carrot, and celery and cover with 2½ quarts of water. Bring to a boil over medium-high heat,

then reduce the heat to maintain a simmer. Skim any scum that rises to the surface. Cook until the broth is flavorful, about 1 hour. Strain the broth through a fine-mesh sieve into a bowl, let it cool completely, and then skim away any fat from the surface. Stir the chicken broth into the bowl of the reserved cooking liquid.

4. Roll out the pasta into sheets until you reach the pasta machine's second-to-thinnest setting. Bring a large pot of salted water to a boil. Prepare a large bowl of ice water. Slip the noodles, two or three at a time, into the boiling water and cook them until they're al dente, 1 minute. Carefully scoop the noodles out of the water with a large wire skimmer and slide them into the ice water to stop the cooking. When they're cool, layer them between clean dish towels until you're ready to assemble the lasagna.

5. In a baking dish that's about 9 × 12 × 3 inches, ladle ½ cup of the broth onto the bottom. Cover with a slightly overlapping layer of cooked noodles, cutting them as needed to fill the gaps. Spread ¾ cup of shredded chicken over the first layer of noodles. Sprinkle with 3 tablespoons of the Parmigiano and then a ladleful of the broth. Add a new layer of noodles, overlapping them slightly.

6. Repeat the layers as instructed above, until all the filling ingredients are used, to make a total of six layers. For the final layer, ladle 1 cup of broth over the top of the noodles and then sprinkle with ¼ cup of Parmigiano. Cover the lasagna and refrigerate for at least 2 hours and up to 4 hours, to let the noodles absorb the broth.

7. Position a rack in the center of the oven and heat the oven to 400°F (200°C).

8. Bake the lasagna until it's bubbling at the edges and browned on top, 45 to 50 minutes. Remove from the oven and let rest for 10 to 15 minutes before serving.

SERVES 6

Pork Medallions with Apples and Caciocavallo

SCALOPPINE DI MAIALE CON MELE E CACIOCAVALLO

Campo Letizia raises wild black pigs that live rummaging throughout the forests behind the farm. The animals' all-natural, acorn-rich diet provides exceptional-tasting meat. Slicing the apples thinly on a mandoline will help them break down in the simmering sauce, thickening it and imparting a delicate sweetness.

12 ounces (340 g) pork tenderloin, trimmed

Kosher salt and freshly ground black pepper

½ cup (62.5 g) all-purpose flour

3 tablespoons (44 ml) extra-virgin olive oil

½ cup (118 ml) white wine

1 cup (240 ml) low-salt chicken broth

1 Golden Delicious apple, peeled, cored, and thinly sliced

4 ounces (115 g) caciocavallo or scamorza cheese, thinly sliced

1. Cut the pork tenderloin into ½-inch-thick slices. Lay each slice between two sheets of plastic wrap and pound out to ¼ inch thick. Season the medallions with salt and pepper and then dredge in the flour, shaking off any excess.

2. Heat the oil in a 12-inch skillet over medium-high heat. Add the pork medallions and cook until lightly golden, 1 to 2 minutes. Flip and cook until golden brown, 1 to 2 minutes. Add the white wine, bring to a boil, reduce the heat to a simmer, and cook until reduced by half. Add the broth and the apple slices and continue to cook until the apples begin to break down, about 8 minutes. Cover each pork medallion with a slice of cheese, cover the pan, and cook until the cheese has melted, 2 to 3 minutes. Season to taste with salt and pepper, and serve on individual plates.

SERVES 4

Crispy Pan-Fried Lamb Chops
COSTOLETTE D'AGNELLO FRITTE

Here lamb chops are left on the bone, pounded thin, breaded, and pan-fried, for an Abruzzese-style Milanese. Coating the chops in the egg mixture and letting them rest in the refrigerator for a few hours before cooking helps to infuse the meat with a hint of lemon.

8 lamb rib chops (about 2 pounds/
   1 kg total), frenched

Kosher salt and freshly ground black
   pepper

2 large eggs

2 teaspoons grated lemon zest

1 cup coarse bread crumbs

3 tablespoons (44 ml) extra-virgin
   olive oil

1 lemon, cut into wedges

1. Put the lamb chops between two pieces of plastic wrap or parchment paper. Using a meat mallet, pound them out to ¼ inch thick. Season with salt and pepper. In a large bowl, beat the eggs with the lemon zest. Add the chops to the bowl and coat with the egg mixture. Cover and refrigerate for at least 30 minutes and up to 4 hours.

2. Put the bread crumbs on a large plate and mix in ½ teaspoon salt and a few grinds of pepper. Coat the lamb chops with the bread crumbs and transfer to a large plate. Heat the olive oil in a 12-inch skillet over medium-high heat until shimmering. Add the lamb chops in batches and cook to a deep golden brown on both sides and medium-rare, 2 to 3 minutes per side. Arrange on a serving platter along with the lemon wedges. Serve immediately.

SERVES 4

Fried Cookies Stuffed with Grape Jelly, Chocolate, and Almonds

CAVICIONETTE

A tender, flaky dough infused with white wine envelops a luscious, dense filling of grape jelly and pureed chocolate and almonds. Once these cookies are fried, they're best eaten right away. The filling oozes goodness as you bite into them.

2⅓ cups (300 g) all-purpose flour

1 teaspoon sugar

½ teaspoon table salt

7 tablespoons (105 ml) water

¼ cup (59 ml) canola oil, plus more for frying

4–5 tablespoons (59 to 73 ml) dry white wine

2 ounces (60 g) bittersweet chocolate, roughly chopped

½ cup almonds

⅓ cup grape jelly

1 tablespoon (15 ml) saba or mosto cotto (or substitute honey)

¼ cup powdered sugar

1. Put the flour, sugar, and salt in a food processor and pulse to combine. Add the water and canola oil and pulse until the mixture resembles coarse meal. Add the wine, 1 tablespoon at a time, and pulse until the dough just starts to come together. Turn the dough out onto a lightly floured work surface and knead until smooth. Wrap in plastic and set aside while you prepare the filling.

2. Put the chocolate and almonds in a food processor and pulse until very finely chopped. Transfer the mixture to a bowl. Stir in the grape jelly and saba and mix until combined.

3. Roll out the dough on a lightly floured surface to ⅛ inch thick. Cut the dough with a 3-inch-round cookie cutter into thirty-five circles, rerolling the dough if needed. Put a rounded tablespoon of filling in the center of each round. Fold the dough in half to form a half-moon shape and pinch the edges together to seal. Refrigerate, covered, for at least 30 minutes and up to 24 hours before frying.

4. Fill a 12-inch skillet (preferably cast iron) with ½ inch of oil. Attach a candy thermometer to the side of the pan, making sure it doesn't touch the bottom. Heat the oil to 365°F (185°C) and fry the cookies in batches until golden brown, 1 to 1½ minutes per side. With a slotted spoon, remove the cookies from the pan and drain on a paper-towel-lined baking sheet. Let the cookies cool and then generously dust with the powdered sugar.

MAKES ABOUT 35 COOKIES

Fried Cookies Stuffed with Grape Jelly, Chocolate, and Almonds

Mosto Cotto

While wine may be one of Italy's most recognized agricultural products, an important by-product from just-pressed grapes is a highly coveted commodity in central and southern Italy, almost more than wine itself. During the harvest, cantinas see people lining up with containers, waiting to take home the unfermented juice, known as mosto in Italian. The mosto must be immediately boiled down, before its sugars begin the early stages of fermentation, to produce the dense, viscous syrup referred to as mosto cotto. During Italy's impoverished past, sugar was an expensive commodity not available to most of the population. Mosto cotto was a popular substitute from the abundant grapes growing throughout the region, and the cooked juice became a staple. While sugar is now readily available, fall months bring families together throughout central and southern Italy to make and bottle their own mosto cotto, for use throughout the year in their family's traditional recipes. At Campo Letizia, Nicola simmers huge vats of grape juice overnight. Their concentrated scent permeates the dining room with a lingering touch of sweetness.

Vineyards and view of Campoletizia

PIETRANTICA

Majella National Park preserves some of Abruzzo's most inspiring mountain scenery. Small stone hamlets pepper the jagged peaks and valleys of a majestic and wild land of raw natural beauty. Shepherds tending their flocks dot the landscape, sharing the park with bears, wolves, deer, and chamois.

Atop a treeless plain in the small village of Decontra, the Sanelli family has forged a living farming their land for generations. Their agriturismo, Pietrantica, run by husband and wife Camillo and Marisa, offers a haven of

unspoiled beauty. Camillo's father, Paolino, is a real-life example of the archetypal Abruzzese mountain shepherd. Spry and possessing a boyish grin, he speaks in a soft and soothing tone full of wisdom and wit. His poetic stories about Abruzzo's past and what life was like in the early 1900s have been documented in books, magazines, and television, as Paolino has become a legend and a voice for preserving Majella's culture.

These days Camillo carries on Paolino's passion for the land, continuing the customs of the farm. While sheep are still raised here, grains have become the main agricultural focus. In addition to farro, Pietrantica grows an antique strand of whole wheat known to flourish only in the mountains of Abruzzo. Processed at a local mill, the flour is dark and rich in nutrients, and an artisan pasta maker shapes and dries the farm's own pasta from it. As mass-produced pasta processed from cheap genetically modified grains fills supermarket shelves, more and more Italians are developing allergies to gluten. Camillo and Marisa take great pride in the fact that their pasta will not affect these people. Paolino also instilled in his son a deep respect for the mountains; in his spare time Camillo works as an expert guide, leading treks and backcountry ski expeditions in the park. An all-day excursion into the heart of the Majella brings famished guests back to the farm, ready to indulge in a healthy and hearty Abruzzese alpine feast.

Pietrantica sign

Marisa cooks from the heart. Her dishes reflect her region's roots and are filling and full of goodness. Guests sit at a long table with Marisa and Camillo and are treated to dishes all made with products from the farm. Polenta from stone-milled farro has a unique nuttiness, enhanced with a wild mushroom ragù topped with shavings of black truffles. Farro stew is enriched with a vegetable puree; chickpeas add a protein kick to the vegetarian dish. Marisa has revived the ancient practice of foraging, and leads groups into the surrounding fields to identify edible weeds and herbs. She boils wild greens, sautés them in garlicky olive oil, and mounds giant spoonfuls over toasted bread. A neighbor provides geese and turkeys, which she stuffs with herbs and roasts over a bed of potatoes. Despising today's modern supermarkets, Marissa has adopted an old-fashioned bartering system with other farmers in the region, trading her surplus of high-quality wheat flour for olive oil and grapes. She also struck a deal with a local baker who, for a few extra kilos of flour, bakes all of Pietrantica's bread. Marisa's revitalization of this time-honored tradition has created a tight-knit community of Abruzzese farmers and agriturismo owners who meet regularly to discuss and share ideas for promoting their region's agriculture. While Paolino's lucid stories paint an idyllic image of the Majella's bucolic past, Marisa is looking to the future to ensure that her family's livelihood will be here for the next generation.

Marisa picking wild chicory

Farro Stew with Butternut Squash Puree

ZUPPA DI FARRO CON CREMA DI ZUCCA

1 cup farro

Kosher salt

2 tablespoons (30 ml) extra-virgin
olive oil

1 yellow onion, cut into fine dice

1 carrot, cut into fine dice

1 stalk celery, cut into fine dice

1 small butternut squash (about
1½ pounds/680 g), peeled, seeded, and cut into ½-inch pieces

1 28-ounce (793 g) can crushed
tomatoes

¼ cup flat-leaf parsley leaves

½ teaspoon chopped rosemary

1 15-ounce (425 g) can chickpeas or
cannellini beans, drained

Freshly ground black pepper

Grated Parmigiano Reggiano, for
serving

A constant on Marissa's menu, farro is served in abundance to guests at Pietran-tica. The recipe is even clipped to the bags of Pietrantica's farro that she sells at the agriturismo. Guests can make their own batch of this nutritious hearty stew and be reminded of their time in the majestic Majella.

1. Soak the farro in cold water for 30 minutes. Drain the farro, rinse, and then drain again. Put the farro in a 4-quart pot, add 8 cups of water, and bring to a boil over medium-high heat. Add ½ tablespoon of salt, reduce the heat to maintain a simmer, and cook until the farro is tender, about 20 minutes. Drain the farro and set aside.

2. Heat the oil in a 6-quart pot over medium heat. Add the onion, carrot, cel-ery, butternut squash, and a pinch of salt. Cook, stirring occasionally, until the vegetables begin to soften, 5 to 7 minutes. Add the crushed tomatoes, parsley, rosemary, and 2½ cups water; continue cooking until the vegeta-bles are very tender, about 15 minutes. Transfer half of the vegetables to a blender and puree until almost smooth. Return the vegetable puree to the pot. Add the farro to the vegetable puree along with the chickpeas and cook until the flavors have melded together, about 10 minutes. Season to taste with salt and pepper.

3. Ladle into individual bowls and serve with the grated cheese.

SERVES 6

Farro Stew with Butternut Squash Puree

Farro Polenta with Mushroom Ragù
POLENTA DI FARRO CON RAGÙ DI FUNGHI

Kosher salt

8 ounces (226 g) farro polenta

3 tablespoons unsalted butter

½ ounce dried porcini mushrooms

2 tablespoons (30 ml) extra-virgin olive oil

2 small cloves garlic, sliced

10 ounces (283 g) cremini mushrooms, trimmed and chopped

8 ounces (226 g) hen of the wood mushrooms, trimmed and sliced

1 teaspoon chopped rosemary

½ cup (118 ml) dry white wine

Freshly ground black pepper

Black truffle shavings (optional)

In the mountain fields of Pietrantica, farro thrives in the cooler alpine climate. Marissa makes a polenta from the finely milled grain and pairs it with a foraged wild mushroom ragù for a rugged Abruzzese dish great after a day spent in the outdoors. Regular cornmeal polenta can be substituted for the farro polenta, but a good source for the grain is www.farawayfoods.com.

1. Bring 5 cups of water to a boil in a heavy-bottomed 6-quart pot over medium-high heat. Add 1 tablespoon of salt and then, in a steady stream, gradually pour in the polenta, whisking constantly to prevent lumping. Reduce the heat so the polenta slowly bubbles and cook, stirring often with a wooden spoon, until tender and creamy, about 25 minutes. Whisk in the butter until melted. Pour the polenta out onto a baking sheet and spread it with a spatula into an even layer. Refrigerate until firm, at least 1 hour.

2. In a small bowl, soak the dried porcini mushrooms in warm water until softened, about 10 minutes. Drain the mushrooms, reserving 1 cup of the liquid. Roughly chop the mushrooms.

3. Position a rack in the center of the oven and heat the oven to 425°F (220°C).

4. Heat the oil in a medium 12-inch skillet over medium heat. Add the garlic and a pinch of salt and cook until golden brown, about 2 minutes. Add the fresh mushrooms, rosemary, and 1 teaspoon salt; cook until the mushrooms start to soften, about 6 minutes. Add the porcinis and cook until the mushrooms are very tender and release their liquid, about 3 to 4 minutes. Add the white wine and reduce by half. Add the reserved mushroom broth and cook until the broth is reduced by half and the flavors have melded, about 5 minutes. Season to taste with salt and pepper.

5. Cut the polenta into 4-inch squares and arrange on a baking sheet. Roast in the oven until warmed through and lightly golden, about 10 minutes. Transfer each polenta square to an individual plate. Top with a spoonful of mushroom ragù, and sprinkle with truffle shavings if desired. Serve immediately.

SERVES 6–8

FARA SAN MARTINO

Majella National Park in the Abruzzo is a land of jagged peaks and deep valleys. The clean, cold springwaters that flow down from these mountains are the key to one of Italy's most treasured culinary products: pasta. The small town of Fara San Martino lies in the heart of the park. Here pasta manufacturers have built their fortunes from the rushing rivers that slice through the Majella. Since the 1800s, dried pasta has been manufactured in Fara San Martino, with DeCecco being one of the first to recognize the area as an ideal location to set up shop, for both its cold springwaters and the region's production of high-quality, high-gluten semolina flour. Today modern high-tech plants and smaller, artisanal factories are all clustered together at the foot of the Majella mountains in Fara San Martino, giving this little-known region of Abruzzo a reputation as the epicenter of Italy's pasta production.

Herb-Stuffed Turkey Thighs with White Wine Sauce

TACCHINO RIPIENO CON SALSA AL VINO BIANCO

While turkey's role on the Italian table is minimal, a poultry farmer down the road from Pietrantica rears the American birds. Marissa likes to de-bone and stuff the thighs with a pungent herb paste that embellishes the juicy meat.

2 skin-on turkey thighs (about 1 pound/453 g each), boned

Kosher salt and freshly ground black pepper

2 tablespoons finely chopped flat-leaf parsley, plus 2 sprigs

1 tablespoon finely chopped rosemary, plus 1 sprig

2 teaspoons finely chopped sage, plus 3 whole leaves

2 cloves garlic

2 tablespoons (30 ml) extra-virgin olive oil

1 medium yellow onion, cut lengthwise into eighths

1 carrot, cut into 2-inch pieces

1 stalk celery, cut into 1-inch pieces

½ cup (118 ml) dry white wine

1. Position a rack in the center of the oven and heat the oven to 425°F (220°C).

2. Lightly season the turkey thighs all over with salt and pepper. In a large mortar, pound the chopped parsley, chopped rosemary, and chopped sage until crushed. Add the garlic and ½ tablespoon salt, and pound until a paste begins to form. Mix in the olive oil.

3. Rub half of the herb mixture over the flesh of one of the turkey thighs. Repeat with the other thigh. Roll each thigh roughly into a cylindrical shape and tie it with two to four loops of twine to secure.

4. Add the onion, carrot, celery, parsley sprigs, rosemary sprig, sage leaves, white wine and ½ cup of water to a 9 × 13-inch roasting pan. Add the turkey thighs to the pan, cover loosely with foil, and roast in the oven for 20 minutes. Remove the foil, baste the turkey thighs with the pan juices, and continue cooking until the thighs register 165°F (74°C) on a meat thermometer, 35 to 45 minutes. Remove the thighs from the pan and set aside on a carving board, tenting loosely with foil.

5. Heat the broiler to high and return the vegetables to the oven. Broil until the vegetables are nicely browned and remove. Transfer the veggies to a platter with a slotted spoon. Strain the juices through a fine-mesh sieve into a serving bowl.

6. Remove the kitchen string from the turkey thighs and carve into slices. Transfer to the platter with the veggies and drizzle with some of the pan juices. Serve the remaining sauce on the side.

SERVES 6

Herb-Stuffed Turkey Thighs

Ginger Cake
TORTA ALLO ZENZERO

Ginger, not a common ingredient in the Italian pantry, made its way into Marissa's repertoire when returning guests brought her some of the root as a gift one year. She became hooked on its sweet and hot tinge, and she uses it to flavor this cake, which she serves at tea—another non-Italian practice she has adopted at Pietrantica.

Softened unsalted butter as needed for pan

4 cups (500 g) unbleached all-purpose flour, plus more for pan

1½ teaspoons baking powder

1 teaspoon baking soda

Pinch of table salt

6 large eggs

1 cup granulated sugar

1 cup (240 ml) extra-virgin olive oil

2-inch piece fresh ginger, peeled and finely grated

1. Position a rack in the center of the oven and heat the oven to 325°F (170°C). Butter and flour a 9-inch bundt pan.

2. In a medium bowl, combine the flour, baking powder, baking soda, and salt.

3. In a stand mixer with a whisk attachment, beat the eggs and sugar together on medium speed until thickened and pale yellow, about 5 minutes. Add a third of the flour mixture to the bowl and mix until just combined. Add half the oil and mix until combined. Alternate adding the flour and oil, ending with the flour and mixing until just combined. Mix in the ginger.

4. Pour the batter into the prepared bundt pan. Bake until golden brown and a toothpick inserted in the center comes out clean, 30 to 35 minutes. Let the cake cool in the pan for 5 minutes, then invert it onto a wire rack to cool completely. Slice and serve.

SERVES 8 TO 10

CHAPTER 7

❧ MOLISE ❧

MASSERIA SANTA LUCIA

The eastern half of Molise consists largely of flat plains that extend down to the coast. Driving west on SS78, which cuts through the region, a drastic change appears in the landscape as you enter the portion known as Alto Molise. Made up of soft rolling hills, mountain chains, rushing rivers, and well-preserved ancient villages, Alto Molise can compete with any region for its inspiring beauty, yet it remains virtually unknown by foreigners

and stands as one of Italy's last undiscovered areas. Masseria Santa Lucia sits in a deep valley here, in the shadows of the small mountaintop medieval village of Agnone. Agnone's wealthy past—it was known for its metalwork in bronze, copper, and gold—is evident in the picturesque and perfectly preserved historic center filled with architecturally inspiring churches and winding cobblestone streets. The surrounding fertile countryside provided abundant quality ingredients.

Here, in a quiet country setting, the Gassano family carries on the rich traditions of their hometown at their country farm and agriturismo, Masseria Santa Lucia. Husband and wife Decio and Emma keep their main residence in town, where Decio has his own accounting firm and Emma once taught high school literature. They have always owned a house in the country, where they would spend the summer and fall months with their children, growing vegetables and making wine and olive oil. Their decision to open an agriturismo came about when Emma retired from teaching, and today the couple divides their time between their urban home and their country farm, spending nearly the entire summer at the agriturismo.

Decio and Emma's unbridled love for the Alto Molise is shared with all who visit the farm. Decio eats meals with guests, whether outside beneath a veranda in the summer or indoors before a cozy fire in the winter,

MOLISE

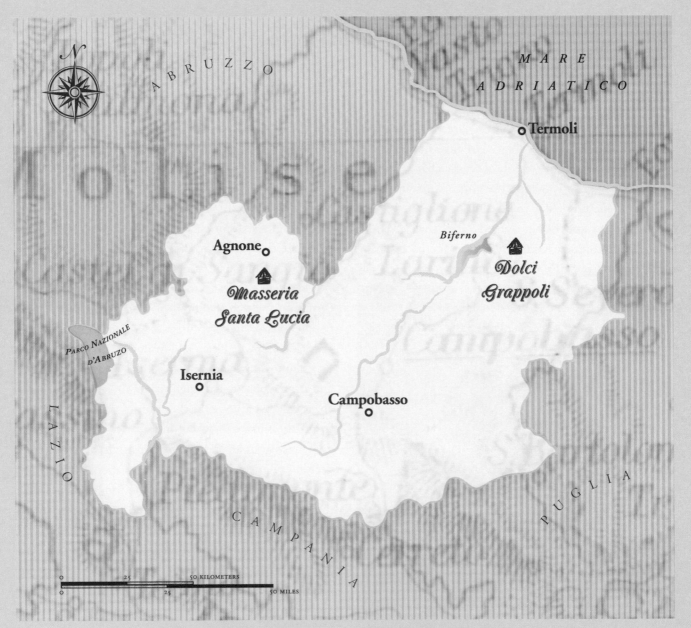

ABRUZZO

MARE
ADRIATICO

○ Termoli

Agnone ○

Biferno

*Masseria
Santa Lucia*

*Dolci
Grappoli*

Parco Nazionale
d'Abruzo

Isernia ○

Campobasso ○

LAZIO

CAMPANIA

PUGLIA

25   50 KILOMETERS
25   50 MILES

and waxes poetic about the treasures of his homeland. A passionate storyteller, he recounts tales of the past—yet conversation always seems to revert to the food Emma prepares. She presents each dish with an expression of nervous exhilaration, which turns to sheer joy when she returns to empty plates and requests for seconds.

Emma's cooking reflects Masseria Santa Lucia's location at the crossroads of central and southern Italy. Every winter she and Decio produce their own salami, pancetta, sausage, and a snow-white rendered pig lard known as strutto. This fat is a key ingredient in many of her dishes, helping build depth and flavor with its mild porkiness. Fresh, handmade pastas are prepared with two types of dough, one in the traditional southern Italian manner with only flour and water, another using eggs to reflect northern influences. First courses include small irregular square-shaped pasta with chickpeas and larger diamond-cut pasta with pork meatballs simmered in tomato sauce. A particular type of cheese known as caciocavallo di Agnone, whose delicate taste derives from the milk of alpine cattle, gets grated into a stuffing for round ravioli that are deep-fried in the farm's own extra-virgin olive oil. While Molise is not typically known for its meat, lamb plays an integral part in its pastoral heritage. Emma slowly braises lamb with parsley, garlic, water, and a spoonful of strutto to create a flavorful sauce for pasta and a fork-tender second course. Meals are always finished with an array of desserts, Emma's true passion, and washed down with one of her homemade liquors from the farm's fruit orchard. With Emma's love for cooking and Decio's profound historical knowledge of Molise, the couple has had great success attracting visitors from all corners of the world. Once they discover the pleasures of Italy's least touristy region, these guests return again and again.

Alto Molise landscape

Fried Spaghetti Nests with Prosciutto and Scamorza

MATASSINE

4 ounces (115 g) spaghetti

Kosher salt

1 ounce (30 g) prosciutto, cut ⅛ inch thick and then into small dice

1 ounce (30 g) scamorza or mozzarella cheese, cut 1/8 inch thick and then into small dice

1 large egg plus 1 yolk, beaten

2 tablespoons grated Parmigiano Reggiano

Freshly ground black pepper

Canola oil, as needed

Frying spaghetti is in every Italian's repertoire as the perfect way to cook up the leftover pasta. After a quick dip in shimmering oil, the strands have a pleasant crunchy bite. Refrigerating the bundles beforehand helps keep them together while they cook.

1. Bring a large pot of well-salted water to a boil. Drop in the pasta and cook until just al dente, according to the package directions. Drain the pasta and then run under cold water and drain again.

2. Put the spaghetti in a large bowl. Add the prosciutto, scamorza, egg, yolk, Parmigiano, a pinch of salt, and a few grinds of pepper. Mix until fully combined. With a fork, pick up some of the spaghetti mixture and then twirl the spaghetti against your hand or a plate to make the nest. Transfer to a large plate and continue making the nests with the remaining mixture. Some of the filling may fall to the bottom of the bowl; once all the nests are shaped, you can tuck the extra filling into each nest. Cover the plate with plastic wrap and refrigerate for at least 30 minutes and up to 2 hours.

3. Heat ½ inch of oil in a large skillet or cast-iron pan over medium heat until shimmering. Place the nests, in batches, in the pan and cook until golden brown, about 2 minutes. Flip the nests over and continue cooking until golden brown, about 2 minutes. With a slotted spoon, transfer the nests to a paper-towel-lined plate and continue frying the remaining nests. Serve immediately.

MAKES 6–8 NESTS

Making Spaghetti Nests

Caseificio Di Nucci

Molise's historical poverty and a scenic landscape of meadows and fields created a shepherding culture based on the milk, food, and wool their animals provided. These elements also created a community of artisanal cheese producers. In the mountain pastures of Alto Molise, the Di Nucci family has been making cheese since 1662, and is recognized as one of the oldest dairies in all of Italy. In the quaint hilltop town of Agnone, Franco Di Nucci represents the tenth generation of his family's cheesemaking legacy. By strictly adhering to his ancestor's time-honored philosophy of making quality cheese by hand, Franco sources only cow's milk from nearby farmers whom he knows and trusts. By modernizing production and sterilization methods, he has made the company a formidable player amongst top producers in Italy, winning top prizes at the country's most respected cheese conference in Lombardy. Di Nucci's success has put Molise on the international gastronomic map as a destination for serious cheese lovers, who visit the shop to buy the hanging gourds of their best-selling caciocavallo.

Caciocavallo plays a formidable role in the region's cuisine, and its versatility in the kitchen is vast. It can be aged from a few months to several years. Younger versions are mild and served as an antipasto with cured pork products or can be used as a melting cheese. The oldest and most expensive cheeses are slightly piquant and grated into fillings for meatballs and involtini, or over soups or pastas. Di Nucci's caciocavallo represents the cheese at its very best, and is appreciated by discerning patrons like Emma and Decio from Masseria Santa Lucia, who serve Di Nucci products to their guests and in their cooking.

Spaghetti with Lamb Reduction
SPAGHETTI CON SUGO D'AGNELLO

The rich lamb stock makes an intensely concentrated sauce packed with good gamey flavor. If you have trouble sourcing lamb bones, any inexpensive cut will do the trick in this dish.

1. Wash the bones well and then put them in a wide 6-quart pot. Fill with enough water to cover by ½ inch. Bring to a boil over medium heat, reduce the heat to maintain a simmer, and skim away any scum that rises to the surface with a slotted spoon. Once no more scum is emitted from the lamb, add the parsley, garlic, butter, and a generous pinch of salt and pepper. Cook at a very gentle simmer until the broth is fragrant and rich, 1½ to 2 hours. Remove the lamb bones from the broth. Bring the broth up to a boil, reduce the temperature, and simmer vigorously until the broth is reduced to a nice thick glaze, 25 to 30 minutes. Season to taste with salt and pepper.

2. Bring a large pot of well-salted water to a boil over high heat. Drop in the pasta and cook until al dente according to the directions on the box. Drain well and then toss to coat with the lamb reduction. Season generously with black pepper and serve with pecorino on the side.

SERVES 4

12 ounces (340 g) lamb bones

¼ cup finely chopped parsley

1 clove garlic, minced

1 tablespoon unsalted butter

Kosher salt and freshly ground black pepper

12 ounces (340 g) spaghetti

Freshly grated pecorino, for serving

Pasta Diamonds with Pork Meatballs

TACCONELLE CON PALLOTTE

1 recipe egg pasta dough (see "Basic Recipes")

2 tablespoons (30 ml) extra-virgin olive oil

1 28-ounce (793 g) can crushed tomatoes

½ small red onion

Kosher salt

1 pound (453 g) ground pork

3 large eggs, lightly beaten

1 cup grated Parmigiano-Reggiano

Freshly ground black pepper

1 cup fresh coarse bread crumbs

Tacconelle are a local pasta made from regular white flour and semolina, shaped into irregular diamonds. At Masseria Santa Lucia, Emma makes a pork meatball tomato sauce to serve with the pasta. Rather than frying the pork balls, she places them into the simmering sauce and slow-cooks them, allowing the sauce to become infused with the meaty flavor.

1. Divide the dough into four pieces. Pass one piece of dough at a time through a pasta machine on the widest setting. Fold the piece of dough over and pass it through the widest setting several more times until the dough is very smooth. Begin reducing the pasta machine setting a notch at a time, and pass the dough through once on each setting, until the dough is ⅛ inch thick.

2. Lay the dough on a flat surface dusted lightly with flour and let dry slightly. It should still be tacky and pliable. Finish with the remaining pieces.

3. With a knife, cut each sheet of pasta lengthwise into 1½–inch-thick strips. Then cut the pasta on a bias crosswise into 1-inch pieces. Transfer to a baking sheet lined with parchment.

4. Start making the sauce. In an 8-quart pot, heat the oil over medium heat. Add the crushed tomatoes, onion, a pinch of salt, and ¼ cup of water. Bring to a boil, cover, and reduce the heat to maintain a gentle simmer.

5. While the sauce is simmering, make the meatballs. In a large bowl, add the pork, eggs, Parmigiano, 1 teaspoon salt, and ½ teaspoon black pepper, and mix well with your hands. Add the bread crumbs and mix until everything is nicely distributed.

6. To shape the meatballs, lightly dampen your hands with cold water and then gently scoop up a handful of meat (about ½ cup) and roll it into a nice even ball. Transfer to a large plate. Continue forming the meatballs, dampening your hands as necessary to keep the mixture from sticking to your hands.

7. Add the meatballs to the simmering sauce and cover the pot. Simmer together until the meat flavor has infused the sauce and the meatballs are cooked through and tender, about 1 hour. Transfer the meatballs to a large serving platter.

8. Meanwhile, bring a large pot of well-salted water to a boil. Drop the pasta into the water and cook until al dente and the pasta floats to the surface, 2 to 3 minutes. Drain the pasta well and then transfer to a bowl. Ladle a few spoonfuls of sauce over the pasta and toss until lightly coated. Serve the pasta with the meatballs and a little extra sauce on the side.

SERVES 6

Emma shaping meatballs

Little Cakes Filled with Ricotta and Chocolate

BOCCONOTTI

Bocconotti are tender little cakes made in several regions of central and southern Italy. Each area has its own version. Emma's are soft and cakey and get filled with sweetened ricotta cream with specks of finely chopped chocolate.

For the cakes:

1 tablespoon unsalted butter, for the pan

2 large eggs

½ cup granulated sugar

½ cup (118 ml) whole milk

½ cup (118 ml) canola oil

1⅓ cups (150 g) all-purpose flour

½ tablespoon grated lemon zest

¾ teaspoon baking powder

½ teaspoon baking soda

For the filling:

1 cup (250 g) ricotta

2 tablespoons granulated sugar

1 tablespoon (15 ml) light rum

2 ounces (60 g) dark chocolate, finely chopped

3 tablespoons powdered sugar

1. **Make the cakes:** Position a rack in the center of the oven and heat the oven to 350°F (180°C). Butter a standard twelve-cup muffin tin.

2. In a stand mixer fitted with a paddle attachment, combine the eggs and the sugar and beat on medium speed until pale yellow and thickened, about 3 minutes. Combine the milk and oil together in a liquid measuring cup. In a medium bowl, combine the flour with the lemon zest, baking powder, and baking soda.

3. Add about one-third of the flour mixture to the stand mixer, beating on medium-low until combined. Add half of the milk, mixing on medium-low until incorporated and scraping the bowl as needed. Mix in another third of the flour mixture, then milk, and then the remaining flour mixture, beating after each addition until just incorporated.

4. Scoop the batter into the prepared muffin tin, filling each cup half way. Bake in the oven until lightly golden brown and the tops spring back when lightly touched, 11 to 13 minutes. Transfer to a wire rack and let cool for 5 minutes. Pop the cakes out of the pan and let cool completely on a rack.

5. **Make the filling:** In a large bowl, whisk together the ricotta and sugar until smooth. Add the rum, and whisk until combined. Gently stir in the chocolate pieces.

6. With a serrated knife, cut off the top of each little cake and set aside on a plate. With a small spoon, scoop out a little of the cake from the center. Spoon 1 heaping tablespoon of ricotta filling into each cake and then put the tops back on. Right before serving, generously dust with powdered sugar.

MAKES 12 CAKES

TRATTURO

The Molise countryside of rolling hills and small, low-lying mountain ranges creates a cool and comfortable summer climate that is lush and verdant. This rich landscape afforded shepherds an ideal environment for their animals to graze on fresh herbs and grasses during the spring and summer months. When the first cold days of autumn arrived, they would lead their animals south, to the warmer plains of sunny Puglia, where they would spend the winter, returning to Molise in the spring. This ritual became known as the Transumanza, and over time a path known as the Tratturo was designated for this procession. Enormously wide and matted down from the trekking of the thousands of animals making their biannual journeys, served as the main artery linking the regions.

Today modern innovations in agriculture have made the Transumanza obsolete, but the Tratturo has been designated a historical landmark, and construction on its tracks is strictly prohibited. Well-marked trails now afford hikers the opportunity to walk the path of the shepherds and their animals, giving a sense of Molise's pastoral past.

I DOLCI GRAPPOLI

With a surname that translates directly as "of the grape," the D'Uva family seemed destined for a life's work of producing wine. For four generations the family has been harvesting grapes in the foothills of Molise, and Angelo D'Uva currently runs the family's cantina with a focus on modern techniques and equipment. This philosophy was instilled in him at an early age by his grandfather, who is recognized as the first in the area to own a press, abandoning the age-old tradition of stomping grapes by feet. Angelo, his sister Teresa, mother Rosa, and father Sebastiano decided to pursue making DOC wines as a family career when Angelo won a local competition for his wines that included a grant to fund professional equipment for a cantina.

The D'Uva winery began with very little money, and the family needed to lease land to expand their vineyard, where rows of Montepulciano, Trebbiano, and the local variety of Tintilla were planted. As construction began and while they were waiting for their saplings to grow into vines, a decision was made to expand their project into a larger operation that included the I Dolci Grappoli agriturismo. Funded by a government grant, the main farmhouse was restored into rooms for overnight guests as well as a restaurant. Over the past decade, Angelo and his father Sebastiano have been slowly establishing a name for themselves among some of Molise's top winemakers. They now own the land they once could not afford to purchase, and the dedication and devotion of the D'Uva family is an inspiring rag-to-riches story. Today they are

proud proprietors of their award-winning winery, and their nectar decorates wine lists as far away as Japan and the United States. In addition to the winery's accolades, Mama Rosa and daughter Teresa's I Dolci Grappoli dining room has been heralded as an exquisite destination to taste traditional dishes of Molise. Their success has brought international recognition to the little-known region as well as a trail of wine-loving travelers who visit the agriturismo for an intoxicating vacation at the foot of the region's softly rolling hills.

Mama Rosa looks like an iconic Italian grandmother, warm and loving and always ready to feed friends and family, and her presence in the kitchen validates I Dolci Grappoli as a special place to experience Old World dishes. Antipasti highlight the variety of cured pork products the

family produces each winter, and a plate of snow-white lardo wrapped around breadsticks always begins meals. During the day, a table in the dining room serves as a stage for the sacred ritual of making pasta. Here Rosa and Teresa hand-shape all of their pastas. Cavatelli, taglierini, and pappardelle are all made with the farm's flour, whose high levels of gluten give it an extra-toothy, al dente texture. Sauces consist of vegetables from I Dolci Grappoli's garden, such as a stewed swiss chard and tomato sauce, as well as meat ragùs and salt cod stewed with tomatoes, onions, and raisins. Another pasta favorite, San Giuseppe, has bucatini tossed with bread crumbs, pine nuts, raisins, bay leaves, almonds, walnuts, cinnamon, nutmeg, and generous grindings of black pepper. Huge wheels of semolina bread are baked weekly in the brick oven; their remains turn into the bread crumbs that make the dish so appealing. Second courses are often grilled and bursting sausages dripping with the juices of well-fed pigs; thick slices of caciocavallo cheese come to the table charred and bubbling. Everything that emerges from the kitchen has been cooked with the farm's own extra-virgin olive oil, and Teresa's degree as a professional olive oil taster ensures quality. Her keen palate allows her to pair D'Uva's wines with each course, further enhancing a dining experience that concludes with a balanced, not-too-sweet white dessert wine matched with a jelly-filled crostata or a local type of fried beignet. The humble cooking of Mama Rosa and the subtle grace of D'Uva's cellar showcase the culinary pleasures that exist in the little-known region of Molise.

Swiss Chard and Caciocavallo Sformato

SFORMATO DI BIETOLE E CACIOCAVALLO

1 large bunch (about 1 pound/
453 g) Swiss chard

3 tablespoons (44 ml) extra-virgin
olive oil

1 clove garlic, smashed

Kosher salt

1 tablespoon fine bread crumbs

1 recipe béchamel (see "Basic
Recipes")

1 cup grated aged caciocavallo
cheese

2 large eggs

Freshly ground black pepper

Caciocavallo, one of Italy's oldest cheeses, traverses the central and southern regions. It falls into the cheesemaking category of "pasta filata," or stretched curd, the same process used for making mozzarella. Cow's-milk curds are heated and shaped into a gourd-like shape and then straddled over a wooden beam to age—hence the name caciocavallo, meaning "cheese on horseback." The cheese can be consumed at various ages. An aged version is ideal in recipes like this sformato, which pairs nicely with the earthiness of the swiss chard. Look for caciocavallo's iconic shape in any specialty Italian market.

1. Cut the stalks of the chard off just below each leaf and thinly slice the stalks. Chop the chard leaves into large pieces.

2. Heat 2 tablespoons of the oil in a large skillet over medium heat. Add the garlic and cook until the oil is fragrant, 1 to 2 minutes. Discard the garlic. Add the swiss chard and a pinch of salt, cover, and cook until the chard is wilted and tender, about 10 minutes. Remove the lid and continue cooking to let some of the moisture evaporate, about 5 minutes. Transfer the chard to a food processor and puree until smooth.

3. Position a rack in the center of the oven and heat the oven to 350°F (180°C). Oil six 4-ounce ramekins with the remaining oil and coat the inside lightly with the bread crumbs.

4. Mix together the pureed swiss chard, béchamel sauce, caciocavallo cheese, eggs, and a few grinds of pepper until combined. Divide the mixture among the prepared ramekins and arrange on a baking sheet. Bake in the oven until puffed up and lightly browned on the top, about 35 to 40 minutes.

5. Remove from the oven and let cool slightly. Invert the ramekins onto individual appetizer plates and carefully remove the ramekin from the sformato. Serve immediately.

SERVES 6

Field of Swiss chard

Cavatelli with Swiss Chard and Tomato Sauce

CAVATELLI CON BIETOLE E POMODORO

2 tablespoons (30 ml) extra-virgin olive oil

1 clove garlic, smashed

1 large bunch swiss chard (about 1 pound/453 g), trimmed and chopped

Kosher salt

1 28-ounce (793 g) can whole peeled plum tomatoes, chopped with their juices

Freshly ground black pepper

1 pound (453 g) cavatelli pasta

Grated Parmigiano Reggiano, for serving

The term cavatelli *refers to a particular pasta shape that is eaten throughout central and southern Italy, especially in Molise and Basilicata. The name translates as "scooped out," and the pasta bears a slight resemblance to gnocchi with a hollowed center. The form is perfect for catching sauces, allowing pasta and sauce to be eaten together in one bite. At Dolci Grappoli, Teresa and her mother hand-roll and -cut their cavatelli each week, transforming their dining room into a pasta-making laboratory. For a quick and delicious weeknight meal, we recommend using store-bought cavatelli for the recipe below.*

1. Heat the oil in a 12-inch skillet over medium heat. Add the garlic and cook until the oil is fragrant and the garlic is golden brown, 1 to 2 minutes. Add the swiss chard to the pan with a pinch of salt, cover, reduce the heat to medium-low, and cook until the chard is wilted, about 5 minutes. Add the tomatoes to the pan and cook, uncovered, until the chard is very tender, 15 to 17 minutes. Season to taste with salt and pepper.

2. Bring a large pot of well-salted water to a boil over high heat. Drop in the cavatelli and cook until al dente, 5 to 7 minutes for fresh pasta or following the direction on the package for boxed. Drain the pasta well and toss with the swiss chard sauce. Transfer to a large platter and serve with the Parmigiano on the side.

SERVES 4 TO 6

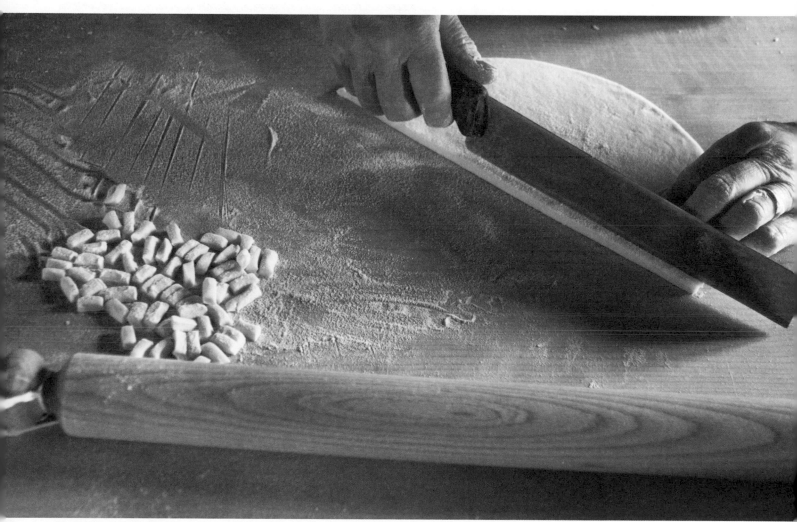

Cavatelli in the making

Bucatini with Fried Bread Crumbs, Pine Nuts, Raisins, and Bay Leaves

MOLLICA DI SAN GIUSEPPE

2 cups coarse bread crumbs

2 cloves garlic, minced

1/3 cup pine nuts

1/4 cup chopped walnuts

1/4 cup chopped almonds

1/4 cup chopped flat-leaf parsley

1/4 cup golden raisins

6 fresh bay leaves, crumbled

1 tablespoon grated orange zest

2 teaspoons freshly ground black pepper

1/4 teaspoon ground cinnamon

1/4 teaspoon freshly ground nutmeg

Pinch of cloves

6 tablespoons (88 ml) extra-virgin olive oil, plus more for serving

1 pound (453 g) bucatini

San Giuseppe is a patron saint of southern Italy, prayed to for all things related to families. Italians celebrate his day on March 19, which is devoted to banquets and feasts. This pasta dates back to the Renaissance times, when sweet and savory ingredients were often combined. The bread crumbs are believed to resemble sawdust, hence their use in this dish paying tribute to Giuseppe—a carpenter by trade.

1. In a large bowl, mix together the bread crumbs, garlic, pine nuts, walnuts, almonds, parsley, raisins, bay leaves, orange zest, black pepper, cinnamon, nutmeg, and cloves.

2. Heat 3 tablespoons of the oil in a large skillet over medium heat. Add half of the bread crumb mixture to the pan and cook, stirring continuously, until the bread crumbs are a deep golden brown, 2 to 3 minutes. Transfer to a plate. Heat the remaining 3 tablespoons of oil in the pan and cook the remaining bread crumb mixture in the same manner. Transfer to the plate.

3. Bring a large pot of well-salted water to a boil over high heat. Drop in the pasta and cook until al dente, according to the instructions on the box. Drain well. Return to the pot and toss with the bread crumb mixture. Divide among individual plates, drizzle with olive oil, and serve.

SERVES 4

Stewed Salt Cod in Tomato Sauce

BACCALÀ CON SALSA DI POMODORO

Although southern Italy is surrounded by waters rich with fresh seafood, salt-preserved cod remains a favorite fish. The flattened, chalky white fillets are piled on top of one another in crates and sold in markets across the southern regions, perfuming the air with a pungent odor. This recipe stews the reconstituted cod with tomatoes. The delectable sauce retains a pleasant fishy taste that will satisfy even those who are not fans of salt cod.

1. One day ahead, rinse the salt cod, put it in a large bowl, and cover with cold water. Cover with plastic and refrigerate for 24 hours, changing the water several times. Drain the salt cod and cut into four 3-ounce portions.

2. Heat the oil in a 10-inch straight-sided sauté pan over medium heat. Add the onions and a pinch of salt and cook, stirring occasionally, until tender and translucent, 5 to 6 minutes. Add the tomatoes, parsley, and ½ cup water and gently simmer for 10 minutes. Add the salt cod to the pan; cover, reduce the heat to medium-low, and cook, flipping the cod halfway through, until cooked through and flaky, 25 to 30 minutes. Stir in the raisins and cook 1 minute more. Transfer to a platter and serve.

SERVES 4

12 ounces (340 g) dried
   salt cod
2 tablespoons (30 ml) extra-virgin
   olive oil
2 small yellow onions, halved and
   cut into ¼-inch wedges
Kosher salt
2 cups canned crushed tomatoes
2 sprigs flat-leaf parsley
¼ cup raisins

Salt cod

Mama Rosa cooking pasta

Throughout the twentieth century southern Italians left their homeland to seek work and opportunity abroad, transplanting their culture around the world. As strangers in foreign lands, they bonded together to create communities and neighborhoods to continue living as they had back home, and the strongest connection with their origins was cooking. The heirloom recipes prepared and shared together in their new lives served as a living lifeline to their heritage. As they began assimilating, starting families, and growing roots in their adopted countries, their food culture became a beloved addition to societies around the globe. Southern Italian cooking became recognized as the pinnacle icon of Italian cuisine.

There exist hundreds, if not thousands, of different shapes of pasta made on well-worn wooden boards in homes throughout the south. Even within each region there exist discrepancies between forms and techniques, but whatever the case, all southern Italian pasta is made with water and durum wheat flour, and is dried before cooking. High-quality semolina becomes the star of perhaps Italy's most famous bread, pane di Altamura from Puglia, whose crackly crust and chewy, light yellow interior are the result of being baked in well-seasoned wood-fired brick ovens. The bread is believed to have origins in the Basilicatan town of Matera, whose inhabitants theorize that the Pugliese stole the secret recipe and have marketed the bread as their own creation. Flour also stars as the south's favorite ingredient in the charred, blistered, and bubbling pizzas of Naples, cherry-tomato-topped focaccia of Puglia, and twice-baked hard biscuits known

as freselle. Bread crumbs are sprinkled over pasta for a nice textural crunch and often favored over cheese.

The southern Italian triumvirate of tomatoes, eggplant, and zucchini decorates menus throughout Campania, Calabria, Basilicata, Puglia, and Sicily, and these regions share a cuisine largely made up of the bounty of fresh vegetables that sprout from fertile soil and sun-filled days. Dispersed among Puglia's olive groves are fields of leafy greens like swiss chard, broccoli rabe, and turnip tops, which are sold in markets across the south; from uncultivated fields sprout wild grape hyacinth that resemble shallots and are known as lampascioni. In addition to producing exceptional oil, the olive trees across the regions provide a plethora of edible olives that are brined and cured and are an indispensable snack for southerners. The town of Tropea in Calabria grows a particular type of purple sweet onions that are heralded even up north, while several varieties of hot peppers satisfy the Calabrese hunger for spice. Basilicatans like things a bit milder and plant a native type of sweet peppers that are dried and hung in kitchens for use throughout the year, and the cooler climate of the region also produces starchy potatoes. Tomatoes will always be Campania's claim to culinary fame, especially those grown in the volcanic soil around Mount Vesuvius. The terraces that cascade down to the sea of the Amalfi coast are home to a variety of the famous lemons that are squeezed to make the tourist-favorite limoncello. Their leaves are used to infuse a touch of their delicate flavor to savory dishes. Citrus is also the backbone of agriculture in Sicily, which is home to a special variety of blood oranges, prized for their delicate sweetness and balanced acidity. Salty capers grow abundantly across the island and also in Puglia, and when combined with olives and preserved anchovies, the trio of ingredients packs a punch of Mediterranean flavor. Together they are a universal component of southern cooking. Almond, walnut, and pistachio trees provide nuts to all of the southern regions and are added to desserts or crushed and sprinkled over dishes. Both cultivated and wild herbs are important components in southern Italian cooking. Bay trees grow everywhere in the south, and their leaves are plucked fresh and added to long-simmering sauces, soups, and stews, and to roasted or grilled meats. Fig trees are another fixture in the southern landscape, their plump green and purple fruit ripe for the taking throughout the summer months. Quince, plums, peaches, and persimmons are other staples that are often cooked down and made into jams.

The fresh cheeses made throughout southern Italy are works of culinary genius, and to experience a freshly made ball of water buffalo mozzarella still warm from the morning's production is an ethereal experience. The same can be said for creamy burrata, billowy soft hand-shaped cow's-milk mozzarella, and just-made ricotta, whose versatility is an essential ingredient in both savory and sweet recipes. Caciocavallo, the south's favorite cheese, is heralded for its melting abilities and can be grilled, eaten on its own, or added to dishes. More piquant aged versions of the cheese are grated over pastas.

While meat does not play as important a role in the cooking of the south, its appearance is omnipresent in the sauces and ragùs that many see as the regions' culinary heart. A pot full of meatballs, braciole, and pork chops simmering away in tomato sauce will forever be one of the classic internationally recognized dishes of Italy. Chickens are reared for their eggs, which are whisked together with cheese and vegetables and baked into frittatas; a traditional peasant dish of Campania poaches them in tomato sauce. The meat is also a popular second course and served roasted with herbs and a squirt of lemon juice. Beef and veal are southern staples that are often finely sliced, stuffed, and rolled into involtini and grilled or roasted. Steaks are thin cuts and commonly get the pizzaiolo treatment, slathered with tomato sauce. The meat is also ground, sometimes with pork, to make meatballs and for a filling for baked stuffed eggplant. Cow stomach lining gets boiled and simmered in broth as a symbol of poorer days when cheaper cuts of meat were all families could afford. Horse is another typical second course of the south, and certain butcher shops are dedicated to its meat. Pigs are raised primarily for salamis and for curing Calabrians add generous amounts of red pepper to theirs. One such, named 'nduja, is a soft spreadable sausage that leaves mouths stained red and tongues burning. Free-range black pigs of the Nebrodi Park in western Sicily have an intensely flavored meat from lives spent feeding on acorn and wild herbs.

The bounty of seafood from southern Italian waters makes for a smattering of dishes. Local fish are grilled and seasoned only with olive oil, salt, and lemon, and taste of the clean salty ocean. Shellfish are highly prized, especially the mussels from Puglia that can be eaten raw, and the small sweet clams of Campania that are steamed open in white wine to make the region's popular linguine with clams. Sea urchins are sliced open, scooped out with a spoon, and eaten while the spines are still moving. Calamari and octopus are boiled to become a base for cold seafood salads or can be braised for hours to create a meltingly tender second course. Migrating tuna pass Sicily's western shores each year, and the fish has become a regional specialty, eaten raw, made into tuna meatballs, and stewed with onions, sugar, and vinegar for a sweet-and-sour preparation.

Southern Italian cuisine speaks of place and origins, born from the waters and sun-kissed land. The generous nature of the southern Italian spirit has infused the cuisine here—the world's favorite comfort food.

Il Giardino di Vigliano farmhouse

CHAPTER 8

CAMPANIA

IL GIARDINO DI VIGLIANO

The Sorrento Peninsula abounds with vibrancy and life, chaos and clamor. Its beauty lies within quaint villages that are set into the cliffs reaching down to the water, with views of Capri, Vesuvius, and the Bay of Naples. Sweet, salty sea breezes mix with bougainvillea and citrus, creating the intoxicating aroma of the Mediterranean. Terraced hillsides with stuccoed villas nestled among olive trees, lemon groves, and brightly blooming flowers make up the idyllic image of Sorrentinian life. It is no wonder thousands of visitors seek out the dolce vita of the Campanian coast each year, descending on the ancient towns, flooding their narrow streets, and choking the roads that wind through the region with gridlock traffic. Above all the madness, tucked away and immersed in a hamlet of gardens and serenity, the Giardino di Vigliano offers respite and an unparalleled way to experience the Gulf of Naples.

Il Giardino di Vigliano began as a silk farm in the early 1300s. Because of its strategic location high above the sea, a tower was erected above the farm's living quarters to be used as a lookout from Turkish invaders. By the mid-1700s the medicinal properties and nutritional benefits of lemons were discovered, especially for sailors on long voyages, and the farm shifted its focus to growing lemons. Today husband and wife Peppino and Ida continue this tradition, and together with their son Luigi run

their cheerful agriturismo, having converted the lookout tower to guest rooms with terraces overlooking the Bay of Naples and Capri.

The lemons of Il Giardino di Vigliano and around the Sorrento Peninsula are of a specific variety, prized for their delicate skin and aroma. Extremely fragrant and succulent, the juice from Sorrento lemons has a delicate acidic tinge with a hint of sweetness. At Il Giardino di Vigliano, they continue to grow and harvest their lemons using an ancient method specific to the peninsula. Tall pergolas constructed of sturdy chestnut wood are fastened together with wire to make an intricate labyrinth above the grove. During the cooler fall and winter months when the trees bear fruit, bamboo mats are placed on top of the poles, blanketing the trees and protecting the lemons. Many farms in the area have stopped using the pergola system, switching to a covering of nylon nets, but Peppino insists that his time-honored way produces more fruit and is better for the health of his trees. Driving down the long, narrow road beneath a canopy of deep green leaves, vivid yellow fruits, and white blossoms stimulates the senses. With the birds chirping and in the lush shade of the garden, the farm has a rain forest feel and seems miles away from the bustle of the street below.

In her basic, no-frills kitchen, Ida cooks Sorrentinian recipes she learned by watching her mother and

grandmother. Her straightforward approach lets the flavors of the food speak for themselves. The rich volcanic soil and salty ocean breezes of the peninsula provide the perfect environment for vegetables to thrive, and these are the base for Ida's dishes. Beneath the vaulted ceilings of the original farm's stables, guests convene at a communal table in the dining room. Ida and Peppino set bowls and plates in the center of the table to be served family-style—as authentic a dining experience as can be found in the area. As the meal progresses, inhibitions are shed, hearts are warmed by good food and wine, and conversations flow. Lemons make frequent appearances at the table, whether slow-cooked with sugar and made into a marmalade for breakfast, sautéed in butter and added to risotto, or infused with liquor for the famed after-dinner shot of Luigi's limoncello. While the ubiquitous spirit has become synonymous with the region, the mass-produced, sickly-sweet sugar-loaded limoncello produced in factories and sold to tourists tastes nothing like the delicate, alcoholic nip Luigi makes from his farm's own lemons.

The hilly land surrounding Il Giardino di Vigliano provides little space for animals, but chickens and rabbits are raised in a pen beneath the lemon trees, and their meat is a common second course. Ida takes advantage of the sea's proximity in many of her dishes, and local fishermen make routine stops at the farm, peddling their catches out of the back of their cars. Ida buys what looks good and is fresh. Native fish make up many of her dishes. A mainstay in her repertoire is a local type of

calamari that she braises with potatoes or slowly simmers with tomatoes for a pasta sauce. Ida, Peppino, and Luigi's Giardino di Vigliano offers genuine home-style cooking nearly impossible to find anywhere else on the peninsula, appealing to those seeking to become a part of the real life of the Sorrento Peninsula, not just another tourist flocking to the same tourist-trap destinations.

Bean and Lentil Soup with Barley

MINESTRA DI FAGIOLI, LENTICCHIE, E ORZO

2 tablespoons (30 ml) extra-virgin olive oil, plus more for serving

2 ounces (60 g) pancetta, cut into fine dice

1 medium yellow onion, cut into fine dice

1 carrot, cut into fine dice

1 stalk celery, cut into fine dice

Kosher salt

1 14-ounce (396 g) can crushed tomatoes

1 cup (6 oz./170 g) black-eyed peas, soaked overnight in cold water and drained

½ cup French lentils

2 sprigs thyme

2 sprigs flat-leaf parsley

1 bay leaf

6 cups (1.5 liters) homemade vegetable broth or canned low-sodium broth

Freshly ground black pepper

¼ cup pearled barley, rinsed

Although it's not much reminiscent of warm and sunny Campania, we were served this soup on a rainy and unseasonably cold summer night, and it felt good eating such a hearty dish. Serve with a slice of grilled Italian bread rubbed with olive oil and garlic.

1. Heat a heavy stockpot over medium heat, add the oil, and cook the pancetta until it begins to render its fat, 2 to 3 minutes. Add the onion, carrot, celery, and ¼ teaspoon of salt; sauté until tender but not browned, 6 to 8 minutes.

2. Add the tomatoes, black-eyed peas, lentils, thyme, parsley and bay leaf to the pot. Cover with the vegetable stock and bring to a boil. Reduce the heat to maintain a gentle simmer and cook until the beans are tender and creamy, 40 to 45 minutes. Discard the thyme, parsley, and bay leaf. Puree half of the soup until smooth, either in a blender or with an immersion blender. Return the soup to the pot and season to taste with salt and pepper. Keep warm.

3. Meanwhile, bring 1½ cups of water to a boil with a pinch of salt and add the barley. Reduce the heat to medium-low and simmer until the barley is tender, about 30 minutes. Drain the barley and then add to the pot with the pureed soup.

4. Ladle the soup into individual soup bowls, drizzle with a little olive oil, and serve.

SERVES 6

LIMONCELLO

No dinner is ever complete on the Sorrento Peninsula or Amalfi coast without a chilled glass of limoncello. Those on the peninsula boast that the liquor they produce is of superior quality than that found on Amalfi due to differences between the lemons used. Those from Sorrento are claimed to be much more fragrant and less astringent. Regardless of your preference, this sweet after-dinner drink is a breeze to make, allowing you to create a bit of sunny Campania right in your own home.

10 large lemons
4 cups (946 ml) 100 proof grain vodka
4 cups (946 ml) water
3 ¾ cups granulated sugar

1. Peel the lemons, making sure not to get any of the white pith (reserve the remainder of the lemons for another use). In a large airtight container, combine the vodka and lemon peel, cover, and let the flavors infuse for 5 days.

2. In a 3-quart pan, add the water and the sugar and bring to a boil over medium heat, stirring occasionally, until the sugar dissolves. Remove from the heat and let cool completely.

3. Add the simple syrup to the infused alcohol and then set aside for 2 more days. Strain the mixture through a fine-mesh sieve into a clean glass bottle, and discard the lemon zest. Serve chilled.

Lemon Risotto

RISOTTO AL LIMONE

1 medium thin-skinned lemon

¼ cup (½ stick) unsalted butter

Kosher salt

1 small shallot, finely chopped

2 cups Arborio rice

½ cup (118 ml) dry white wine

1 quart (946 ml) homemade or low-
    sodium canned vegetable broth,
    warmed

¼ cup freshly grated Parmigiano
    Reggiano cheese

At Giardino di Vigliano this very simple risotto brings out the delicate essence of the Sorrento lemon, whose aromatic thin skin serves as the base to many of the dishes prepared in Ida's kitchen.

1. With a sharp paring knife, cut the lemon peel away from the flesh into 2-inch-thick strips. Remove all of the white pith from the peel. Thinly slice the peels crosswise.

2. In a 10-inch straight-sided pan, melt the butter over medium-low heat. Add the lemon zest and a pinch of salt and cook until tender, about 3 minutes. Add the shallot with another pinch of salt and continue cooking until the shallot is tender and the butter is very fragrant, 2 to 3 minutes. Add the rice to the pan, stir to coat, and cook until lightly toasted, about 2 minutes. Add the wine to the pan and reduce to dry, 3 to 5 minutes.

3. One ladleful at a time, add the broth to the pan to cover the rice, stirring constantly with a wooden spoon until the broth has been completely absorbed by the rice. Continue to ladle broth into the risotto in the same manner, stirring continuously, until the rice is al dente and creamy, 15 to 20 minutes. Stir in the grated Parmigiano and season to taste with salt. Serve in individual shallow bowls.

SERVES 4

Lemon Risotto

Meyer Lemon Marmalade
MARMELLATA DI LIMONE

7 Meyer lemons (2 pounds/about 900 g)

2 cups granulated sugar

Homemade marmalades with farm-fresh fruit are the essence of the Italian agri-turismo, and nearly every farm we visited made its own jams and jellies to serve at breakfast. This was one of our favorites. Its unique, slightly bitter aftertaste is delicious on warm bread drizzled with olive oil or soft butter. It is also delicious paired with sharp cheese, crackers, and a sparkling white wine.

1. Trim the ends from the lemons. Cut 4 of the lemons into eighths lengthwise, remove the seeds, and then cut crosswise into thin slices. Trim away the peel from 2 of the lemons and cut the segments free from the membranes. Juice the remaining lemon to get ¼ cup of juice.

2. Bring a small pot of water to a boil over medium-high heat. Add the sliced lemons with the rind to the water, return to a boil, and cook to remove some of the bitterness, 2 to 3 minutes. Drain well.

3. In a small saucepan, add 1 cup water, the ¼ cup lemon juice, and the sugar. Bring to a boil, stirring until the sugar dissolves, 2 to 3 minutes. Add the lemons and return to a boil. Reduce the heat to low, skim any foam from the surface, and simmer very gently until the mixture breaks down and thickens, 40 minutes to 1 hour, checking frequently to prevent overcooking.

4. To test for doneness, chill a small dish in the refrigerator. Put a small dollop of marmalade on the dish, let it cool briefly, and then run your finger through it. If the mark stays, the marmalade is ready; if it doesn't, cook the mixture for a few more minutes and retest. Cool completely and refrigerate for up to 2 weeks.

MAKES 1½ CUPS

Meyer Lemon Marmalade

Braised Calamari with Parsley and Potatoes

CALAMARI BRASATI CON PREZZEMOLO E PATATE

6 tablespoons (88ml) extra-virgin olive oil

1 small yellow onion, finely chopped

1 small garlic clove, thinly sliced

3 tablespoons finely chopped flat-leaf parsley

1/4 teaspoon crushed red pepper flakes

Kosher salt

1 pound (453 g) calamari, cut crosswise into 1/4-inch-thick slices

3/4 cup (175 ml) dry white wine

1 medium red potato, peeled and cut into 1/2-inch dice

1 lemon, cut into wedges

The terraced hills of the Sorrento Peninsula cascade down to the sea, offering very little land to raise animals. Thus seafood is an integral part of the cuisine. At Il Giardino di Vigliano fresh fish makes more than a few appearances on the menu. Here is one of the dishes that we loved the most: The calamari becomes extremely tender, and the potatoes help sop up all of the flavorful juices.

1. Heat the olive oil in a 12-inch straight-sided pan over medium heat. Add the onion, garlic, parsley, pepper flakes, and a pinch of salt. Cook until the onions are tender and just beginning to brown, 4 to 6 minutes. Add the calamari and cook until lightly browned, 3 to 4 more minutes. Add the white wine, bring to a boil, press a piece of heavy-duty foil down onto the calamari, and then cover with a lid. Reduce the heat to low to maintain a very gentle simmer, and cook until the calamari is tender, 45 to 50 minutes.

2. Add the potato and a pinch of salt, cover, and cook until tender when pierced with a fork and the calamari is meltingly tender, about 10 to 12 minutes. Serve on a platter with the lemon wedges alongside.

SERVES 4

Baked Stuffed Baby Eggplant
MELANZANE RIPIENE

The ubiquitous purple eggplant is synonymous with the Campanian table. Ida stuffs smaller baby eggplant with a beef-and-tomato filling for this satisfying second course.

1. Fill a large bowl with cold water and 1 tablespoon of salt. Cut out the inside flesh from the eggplant, leaving enough intact so each retains its shape. Finely chop the filling and transfer to the bowl; add the hollowed-out baby eggplant and let sit for about half an hour to remove some of the bitterness. Drain the eggplant and pat the halves dry with a towel. Squeeze out any excess water from the chopped eggplant.

2. Position a rack in the center of the oven and heat the oven to 375°F (190°C).

3. Heat the oil in a 12-inch skillet over medium heat. Add the chopped eggplant, 4 tablespoons of the parsley, and a pinch of salt; cook until tender, 3 to 4 minutes. Add the ground beef, ¾ cup of the tomatoes, and the garlic. Continue to cook until the beef is browned and the tomatoes begin to break down, 6 to 8 minutes. Stir in the grated Parmigiano and then remove from the heat. Season to taste with salt. Let cool slightly, then add the egg and stir well to combine.

4. Divide the filling among the hollowed-out baby eggplant. Top each eggplant with the remaining pieces of the tomatoes and sprinkle with the remaining parsley. Arrange the eggplant in a 9 × 13 × 2-inch baking dish and fill the bottom with ¾ cup water. Bake in the oven until the tops are nicely browned and the eggplant is tender when pierced with a fork, 40 to 45 minutes. Remove from the oven and let cool slightly before serving.

SERVES 6

3 baby eggplant (1 pound/453 g total), cut in half lengthwise

Kosher salt

2 tablespoons (30 ml) extra-virgin olive oil

5 tablespoons chopped flat-leaf parsley, plus more for garnish

12 ounces (340 g) ground beef

1 cup grape tomatoes, finely chopped

1 garlic clove, finely chopped

½ cup grated Parmigiano Reggiano

1 large egg

Squeezing excess water from eggplant

IL CORTILE

In the urban commotion that engulfs Naples's suburbs, the Neapolitan zest for life resonates. Streets are alive with the colors and personalities of Italy's most vivacious populace. As more people move away from the heart of the city, these small towns are becoming the essence of true Neapolitan culture. Cicciano lies about twenty miles from the metropolis, and urban sprawl has transformed the once peaceful countryside community into a city that bustles with activity. The present day reveals little of its past as a country oasis for city dwellers, with manor homes surrounded by farms and gardens.

On one of the main streets running off the central piazza, two giant brown doors are the gateway into the Il Cortile agriturismo, and a step back in time. A walk down a stone-vaulted passageway has an Alice-stepping-into-Wonderland feeling, with a magnificent open-air courtyard of citrus trees heavy with lemons, grapefruits, mandarins, and oranges, and flowering vines that crawl up the surrounding buildings. Down another corridor and past iron gates you'll find a verdant green Neapolitan garden, its towering palm and magnolia trees offering respite from the sweltering summer heat.

Arturo Nucci and his wife, Sietske, moved to his family's country estate shortly after being married to live in a more tranquil setting away from Naples. In various greenhouses around Cicciano, they began cultivating decorative houseplants as a business; this still serves as the main agricultural production of the farm. When their three daughters Giovanna, Alessandra, and Ada finished their schooling, they decided to work together as a family and open the doors of their property as the area's first urban agriturismo.

Comfortable and intimate, the farm has only four bedrooms, allowing for a peaceful holiday in the luxurious setting of an old Neapolitan manor house. Inside, the historic architectural structure impresses with massively thick two-foot-wide arches that span the lower level and a narrow staircase that descends into blackness and into the original cantina. When construction first began in the 1600s, the cantina was dug by excavating tufa rock, which was then used for building the base, walls, and arches of the house. The enormous stone basement was cool, dark, and the perfect place for storing food and wine, and the family would make an income renting space to local neighbors to store their goods. On the main floor an elegant dining room offers a refined dining experience, and in the kitchen—a small, comfortable space lined with 1970s retro-tiled walls—Sietske and her daughters feel right at home, cooking up an elevated style of cuisine.

A native of Holland, Sietske bases her dishes largely around recipes she learned spanning her thirty years in Naples. While some of her repertoire represent the simplistic style of her adopted region's flavors, their impeccable execution has made a name for Il Cortile, and has locals booking often for lunch and dinner. Olives, capers, and anchovies, the jewels of Naples cuisine, are omnipresent in many preparations. The three staples melt together in Il Cortile's escarole pie, which sings with salty

briny flavor. The year-round abundance of the agriturismo's fresh citrus has Sietske looking for creative ways to implement their soft acidic tinge. Thin segments of oranges make their way into a shaved fennel salad, thick slices of scamorza cheese are grilled between its leaves, orange juice serves as a base for gelato, and orange zest is mixed into a chocolate cake. A Saturday weekly market brings the abundance of the nearby Tyrrhenian Sea to Cicciano, and unlike most landlocked agriturismi, Il Cortile serves fish as a staple; it's a perfect marriage for the farm's lemons, in dishes such as gratinéed anchovies bathed in the juice.

In a country not known for its breakfast tradition, Il Cortile's table provides an exceptional breakfast that always includes freshly squeezed mandarin and orange juice, all-natural jams made from the farm's own fruit, homemade sweet cakes, and rich Neapolitan coffee. All are served in Il Cortile's elegant dining room on antique linens sought out with a keen eye at area outdoor markets. Since her arrival in Campania in the early 1970s, Sietske has witnessed drastic changes brought about by corrupt politics, overpopulation, and mass tourism, yet she retains an unbridled enthusiasm for the hidden beauty of the region. In a Campania where it's getting more and more difficult to escape the maddening crowd, she sets itineraries for her guests that lead down less-trodden roads and always return to the refined urban agriturismo of Il Cortile.

Griddled Scamorza and Lemon Leaf Bundles

SCAMORZA ALLA PIASTRA AL PROFUMO DI LIMONE

4 ½-inch-thick rounds of scamorza cheese

8 large organic lemon leaves

When Sietske showed us this dish, we really took to its simplicity and decorative, rustic presentation and loved the hint of lemon flavor that the leaves infuse into the melting cheese.

1. Arrange 4 lemon leaves, shiny-side down, on a clean surface. Top each leaf with 1 slice of scamorza cheese and then top each with the remaining leaves matte-side to cheese.

2. Set a griddle or a 12-inch cast-iron skillet over medium heat. When it is hot, carefully place the lemon leaf bundles down on the griddle. Cook until the leaves are fragrant and the cheese on the bottom is softened, 1 to 2 minutes. With a spatula, carefully flip the bundles over and continue cooking until the cheese is very soft, 1 to 2 minutes more. Serve immediately.

SERVES 4

Cauliflower Salad with Orange and Olives
CAVOLOFIORI ALL'INSALATA

In this dish refreshing citrus mellows the strong anchovies and capers, whose flavors add a subtle brininess reminiscent of the essence of the Neapolitan kitchen.

1. Trim away the peel from the oranges and cut the segments free from the membranes, letting them fall into a small bowl. Squeeze the juice from the membranes into the bowl.

2. Bring a 4-quart pot of salted water to a boil over medium-high heat. Add the cauliflower florets and cook until tender, about 8 minutes. Drain the cauliflower well and transfer to a large bowl. Toss with the oil, anchovies, and capers. Then stir in the olives and orange segments along with their juices. Season to taste with salt and pepper.

SERVES 4

2 medium navel oranges

1 head cauliflower (about 2 pounds/ 1 kg), cored and broken into 1-inch florets

2 tablespoons (30 ml) extra-virgin olive oil

2 dry salted anchovies, rinsed, patted dry, and chopped

2 teaspoons salted capers, rinsed and patted dry

¼ cup Gaeta olives (or substitute kalamata olives), pitted and halved

Kosher salt and freshly ground black pepper

Cauliflower in morning light

Escarole Pie

PIZZA DI SCAROLA

Gaeta olives are native to Campania. They are medium-size black olives, not too robust in flavor. This pie is traditionally served on Christmas Eve. In and around Naples, Christmas Eve is a very important evening where families gather to a feast of fish. Since everyone knows what is in store for them that night, they eat this very tasty escarole dish for lunch, because it's light. The dish is so yummy, though, that it has made its way into the Neapolitan diet, no longer being reserved just for the day before Christmas.

For the dough:

2¼ cups (280 g) all-purpose flour

Pinch of table salt

1 cup (2 sticks) unsalted butter, cold, cut into pieces (preferably European-style butter)

¼–⅓ (60–80 ml) cup ice water

For the filling:

2 large heads escarole, washed, trimmed, and torn into large pieces

¼ cup raisins

3 tablespoons (44 ml) extra-virgin olive oil

2 salt-packed anchovy fillets, rinsed well and chopped

2 small cloves garlic, smashed

½ cup pitted Gaeta olives

¼ cup pine nuts

1 tablespoon salt-packed capers, rinsed well

Kosher salt and freshly ground black pepper

1. **Make the dough:** In a stand mixer fitted with the paddle attachment, combine the flour, salt, and butter. Mix on low speed until the dough forms coarse crumbs the size of hazelnuts. Gradually add the water in a steady stream, mixing until the dough just comes together. Turn the dough out onto a clean work surface. Divide it in half, and shape the halves into disks. Wrap the disks separately in plastic and refrigerate for at least 1 hour and up to 1 day.

2. **Make the filling:** Put the escarole in a 12-inch straight-sided sauté pan with 2 tablespoons of water, over medium heat. Cover the pan and cook the escarole, stirring occasionally, until wilted, 8 to 10 minutes. Transfer the escarole to a colander set over a large bowl and let cool. Once it's cool enough to handle, squeeze out the excess water and set aside on a plate.

3. In a small bowl, soak the raisins in 3 tablespoons of warm water to soften, about 5 minutes. Drain the raisins and set aside.

4. Heat the olive oil, anchovies, and garlic in a 12-inch skillet over medium heat until the garlic is golden brown, 2 to 3 minutes. Discard the garlic. Add the escarole, olives, pine nuts, raisins, and capers; cook for 1 minute. Season to taste with salt and pepper. Transfer to a bowl and let cool.

5. **To assemble,** position a rack in the center of the oven and heat the oven to 400°F (200°C). Butter a 9-inch pie plate.

6. If the dough was refrigerated for several hours or overnight, let it sit at room temperature until pliable, about 20 minutes. On a lightly floured surface, roll out one of the dough disks into ⅛-inch-thick circle, 13 inches in diameter, and transfer it to the prepared pie dish. Prick the dough all over with a fork and refrigerate until firm, 15 to 20 minutes. Pour the filling into the pie shell, spreading it out evenly with a spatula.

7. Roll out the second disk of dough as above and set it over the filling to form a top crust. Press the edges of the dough together to seal the crust; trim the overhang to ¼ inch and fold it under. Crimp the dough all around. Using your finger, make a ½-inch circle in the middle of the top crust. Bake in the oven until the crust is lightly golden and set, about 1 hour. Let cool on a wire rack for 30 to 40 minutes. Cut into wedges and serve.

SERVES 6–8

Lettuce fields Campania

Pasta with Squash, Cooked Risotto-Style

PASTA E ZUCCA

This unique method of cooking dry pasta directly in its sauce has broth continuously added as the pasta absorbs the liquid from the simmering sauce. The finished dish has a consistency that is not too soupy but not too dense and similar in body to a risotto. The intensely flavorful pasta is creamy, toothsome, and downright delicious.

1. In an 8-quart saucepan, heat the oil with the garlic over medium heat. Cook until the garlic is lightly golden brown, about 2 minutes. Discard the garlic. Add the squash, bay leaves, pepper flakes, and ½ teaspoon salt, stir to coat with the oil, and cook for 2 minutes. Add enough broth to just cover the squash (about 3 cups). Bring the mixture to a vigorous simmer, reduce the heat, and cover. Cook until the squash is just tender, 10 to 12 minutes.

2. Add the pasta to the pan along with 1 teaspoon salt and cook, adding more broth a ladleful at a time and stirring often, until the pasta is al dente, the squash has broken down, and the sauce is creamy, about 12 to 14 minutes. Season to taste with salt and pepper. Sprinkle with the chopped parsley and serve with grated pecorino on the side.

SERVES 4–5

3 tablespoons (44 ml) extra-virgin olive oil

2 small garlic cloves, smashed

1 medium acorn squash (about 2 pounds/1 kg), peeled, seeded, and cut into ½-inch pieces

2 bay leaves, preferably fresh

Pinch of crushed red pepper flakes

Kosher salt

5½ cups (1.3 liters) vegetable or chicken broth (homemade or canned low-sodium)

12 ounces (340 g) fusilli pasta

Freshly ground black pepper

½ tablespoon chopped flat-leaf parsley

Freshly grated pecorino cheese, for serving

Gratinéed Fresh Anchovies with Lemon
ALICI IN TORTIERA

2½ pounds (1.13 kg) fresh anchovies

5 tablespoons (73 ml) extra-virgin olive oil

Kosher salt and freshly ground black pepper

¼ cup (59 ml) lemon juice

2 large cloves garlic, minced

¼ cup chopped flat-leaf parsley

5 tablespoons bread crumbs

Salty, sweet, delicate, briny, and citrusy, this dish packs a punch of flavors. Fresh anchovies are high in omega-3, contain little or no mercury, and thrive in the waters surrounding Italy, making the small fish a staple of the Mediterranean diet.

1. Position a rack in the center of the oven and heat the oven to 350°F (180°C).

2. Clean the anchovies by removing their heads and tails with a small paring knife. Then split open the bellies and remove the intestines and spines. Finally, remove the small fin on the back (skin side) of the fish. Rinse the anchovies, drain in a colander, and pat dry with paper towels.

3. Drizzle 1 tablespoon of the olive oil on the bottom of a 9 × 13-inch baking dish. Take one anchovy and arrange it flat, skin-side up, in the baking dish. Keep arranging the anchovies in this manner until you have an even layer (you should have used one-quarter of the anchovies). Sprinkle the anchovies lightly with salt and pepper. Drizzle with 1 tablespoon olive oil, 1 tablespoon lemon juice, a pinch of the minced garlic, 1 tablespoon parsley, and 1 tablespoon bread crumbs. Repeat three more times with the remaining anchovies and other ingredients. With the final layer, use 2 tablespoons of bread crumbs on the top.

4. Bake in the oven until the anchovies are flaky and white, 20 to 25 minutes.

SERVES 4–6

COLATURA DI ALICI

Schools of anchovies, locally called pesci azzurri, flourish in the waters around the Bay of Naples and the Amalfi coast. They're caught with nets and sold in markets throughout the country. The town of Cetara on the Amalfi coast produces a pungent aromatic sauce from the oily fish known as Colatura di Alici, made from the extracted juice of salt-preserved anchovies. Rows of anchovies are layered with salt in a ceramic urn set with a heavy weight on top of them, and are left until they begin to release their juices. The amber-hued liquid is drained and filtered, and is then ready for cooking. Colatura di Alici bears a resemblance to fish sauces common in Thai and Vietnamese cooking, and this assertive yet subtle essence of anchovy has made its way into traditional Campanian cuisine. It's commonly paired with other Neapolitan favorites like olives and capers for a pasta sauce or as a simple accompaniment to sautéed vegetables.

Orange Ice Cream with Orange Caramel Sauce

GELATO ALL'ARANCIA

For the ice cream:

3 medium navel oranges

4 cups (960 ml) whole milk

1 cup granulated sugar

Pinch of table salt

8 egg yolks

For the caramel sauce:

1 orange

1 cup granulated sugar

2 tablespoons (¼ stick) unsalted butter

2 tablespoons (30 ml) Cointreau or Grand Marnier

1 tablespoon (15 ml) brandy

½ tablespoon maraschino liqueur

Sietske has mastered the art of making ice cream, and it has become a favorite among her guests. She brought this recipe with her from her homeland of Holland, and uses it with the sweet oranges that grow in her courtyard. She jokes that in Holland this would never taste as good because of the lack of quality citrus there. Fortunately for us in the States we have a great supply of local oranges that can re-create the flavors of Campania.

1. **Make the ice cream:** Finely grate the zest from the oranges. Then juice 1 or 2 of the oranges to get ¾ cup juice. Strain the juice through a fine-mesh sieve into a small bowl and set aside.

2. In a medium saucepan, mix the milk, sugar, and salt. Warm the mixture over medium-high heat, stirring occasionally, until the sugar dissolves and the milk is steaming, 2 to 3 minutes. Remove from the heat and stir in the orange zest. Cover the pan and let it sit for 30 minutes to infuse the flavor.

3. In a large bowl, whisk the egg yolks. Rewarm the milk mixture over medium-high heat until the milk is steaming, 1 to 2 minutes. In a steady stream, pour the warm milk mixture into the egg yolks, whisking constantly to prevent the eggs from curdling.

4. Pour the egg mixture back into the saucepan and cook over low heat, stirring constantly with a wooden spoon, until the custard thickens enough to coat the back of the spoon, 4 to 8 minutes. Strain the custard through a sieve into a medium bowl. Press on the orange zest in the strainer with the spoon to extract as much flavor as possible. Set the bowl into a larger bowl filled with ice water.

5. Cool the custard by stirring it over the ice bath. Stir in ½ cup of the orange juice and refrigerate the custard until completely chilled. Freeze the custard in an ice cream maker. Transfer the ice cream to an airtight container and freeze.

6. **Make the caramel sauce:** Cut the orange into eighths lengthwise. With a paring knife, cut the peel away from the flesh of the orange. Remove any white pith from the peel. Thinly slice the peel. Bring a 2-quart pot of water to a boil over medium-high heat. Add the orange peel and blanch to remove some of the bitterness, about 5 minutes. Drain the peel and set aside to dry on a kitchen towel.

7. Put the sugar and ¼ cup of water in a small pan over medium-high heat. Cook the sugar, stirring, until the sugar dissolves. Bring to a boil and cook until it becomes a deep amber color, 5 to 8 minutes. Remove from the heat and carefully stir in the remaining ¼ cup orange juice, the butter, Cointreau, brandy, and maraschino. Be careful: The mixture may sputter. Return to the heat to smooth out the caramel, then remove from the heat and stir in the orange peels. Pour into a heatproof container and cool completely.

8. **To serve,** scoop the ice cream into a small dish and drizzle with some of the caramel sauce.

SERVES 10

PORTA SIRENA

The Sele plains are an immense stretch of flatlands that skirt the coastline south of Salerno. Covered by the Tyrrhenian Sea thousands of years ago, the area was left an unforgiving swamp as the ocean receded, infested with snakes, scorpions, and mosquitoes. Mussolini had the spot drained, constructing a water and irrigation system to turn the wet ground into tillable farmland. The land proved to be extremely fertile, rich in calcium and minerals left behind from the ocean. Since the land was still laden with moisture, habitants turned to the stout, water buffalo to help cultivate the soggy soil. Soon the high-fat, protein-rich milk of the animals was discovered, and southern Italy's most famous cheese was born. As tractors were introduced to work the farmland of the plains, water buffalo have remained the region's most prized commodity, raised for the production of decadently rich globes of Mozzarella di Bufala. The worldwide success of the cheese brought an influx of farmers seeking to strike it rich with water buffalo farms, and land became scarce, leading the Italian legislature to mandate that one hectare (about two and a half acres) of land was needed for every buffalo. Today the herds of massive free-range horned beasts roaming the plains have vanished, with the majority of the animals living in open-air barns.

Since the early 1930s the Jemma family has been rearing the docile animals outside of the ancient walls of Paestum's UNESCO-protected Greek temples and ruins. Among the first large-scale farming families in the area, the Jemmas are some of the largest landholders in the region.

Their agriturismo, Porta Sirena—named for one of the four original arched doorways entering the historic city—is the latest addition to their farm; the original stables were converted into bedrooms and an elegant restaurant.

The Jemmas' buffalo have been relocated to another property close to both the sea and the family dairy, where fresh milk is turned into mozzarella and sold daily. Before each milking, the buffalo enter a shower room, where they are hosed down, cooled, and washed free of any dirt and manure, assuring a sanitized product ready for transport to the dairy. The Jemmas are also among the first in the Sele to recognize the high quality of water buffalo meat. The older animals do provide a tough and tasteless poor man's meat, but Porta Sirena raises calves, whose steaks can rival the best cuts from Italy's most cherished cattle. This healthy meat has little fat and cholesterol, yet its benefits have seen only slow acceptance into the Italian diet. Porta Sirena has proved to be the perfect venue for showcasing the delicious characteristics typical of well-bred, well-fed water buffalo.

Generous, 2-inch-thick, grilled T-bone buffalo steaks emerge from Porta Sirena's kitchen served sizzling on volcanic rock hot plates. The farm's signature dish needs little manipulation other than a quick sear over hot coals, a sprinkle of salt, and a drizzle of extra-virgin olive oil. With practically no fat content, the meat has a pleasantly tender texture full of flavor from the animals' diet of natural grasses and herbs from the Sele plains. Creamy buffalo mozzarella also plays heavily into the cuisine. Eaten on its own or paired with prosciutto from Cilento Park, whose mountains stand directly behind the agriturismo, the cheese's billowy, mouth-coating texture and salty-sweet flavor are an epiphany of decadence. When rolled with grilled eggplant and baked in tomato sauce, the oozing and bubbling mozzarella creates a harmonious marriage of Campanian flavors. Braised cabbage with stale bread, chicory, and local baby artichokes from Paestum are a few of the seasonal vegetable dishes that make up the constantly changing list of starters, along with fried eggs whose creamy, runny yolks mingle with a fiery cherry tomato sauce for an ethereal taste of utter simplicity. Pasta reflects the array of shapes produced throughout the region—especially paccheri, one of Campania's favorite

shapes—and are sauced with seasonal products and made creamy with generous dollops of buffalo ricotta. Other second courses include house-made buffalo sausages as well as the farm's own chicken. Peach trees surround Porta Sirena, and the kitchen conserves their fruit in marmalades that are used year-round in desserts. A well-appointed wine list showcases robust Campanian vintages well suited for drinking with the hearty, meat-based fare of Porta Sirena. As the popularity of the restaurant grows, so does that of the water buffalo, and Porta Sirena has proved to many in the area its versatility and the overall deliciousness this animal has to offer—far beyond just the making of cheese.

Water buffalo steak

Fried Egg in Spicy Cherry Tomato Sauce
UOVA IN PURGATORIA

3 tablespoons (44 ml) extra-virgin
   olive oil
1 clove garlic, smashed
Pinch of crushed red pepper flakes
1 14-ounce (396 g) can cherry
   tomatoes
Kosher salt
4 extra-large eggs

Italian peasant food at its finest. Purgatoria refers to the spicy tomato sauce. The secret is not to overcook the eggs—you want the runny yolk to mix into the sauce. This is a great appetizer but also makes for a southern-Italian-style American brunch dish.

1. In a 3-quart saucepan, heat 2 tablespoons of the olive oil over medium heat. Add the garlic and cook until fragrant, about 1 minute. Add the crushed red pepper flakes and cook a few seconds. Add the cherry tomatoes and a pinch of salt and cook at a gentle simmer for 25 minutes. Whisk the sauce to break down the tomatoes and garlic; season to taste with salt.

2. In a 10-inch nonstick skillet, heat the remaining 1 tablespoon of oil over medium-high heat. Crack the eggs into the skillet and season with a pinch of salt. Cook until the whites just start to set, 1 to 2 minutes. Ladle a few spoonfuls of sauce over the eggs and cook until the whites are fully set but the yolks are still runny, 2 to 3 minutes. Serve immediately with extra sauce on the side.

SERVES 4

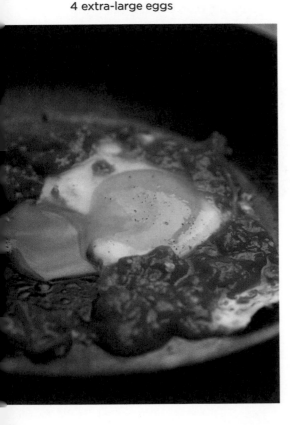

Uova in Purgatoria

Cooked Bread with Savoy Cabbage

PANE COTTO CON VERZA

This robust dish should be made with good crusty Italian bread that is stale and dry. As the cabbage and bread cook together, the bread breaks down, creating a thick and hearty winter meal that is Italian comfort food at its best. Another common preparation uses escarole in lieu of cabbage.

1. Heat the oil in a large straight-sided skillet over medium heat. Add the garlic and cook until fragrant, about 1 minute. Discard the garlic. Add the cabbage and a pinch of salt, cover, and cook, stirring occasionally, until tender, 10 to 12 minutes. Add the broth and tomatoes; continue to cook, covered, until the cabbage is very tender, 20 to 25 minutes.

2. Stir in the bread and Parmigiano and cook until the bread softens, but sill retains some of its texture, 10 to 15 minutes. Season to taste with salt and pepper. Transfer to a platter and serve.

SERVES 6

2 tablespoons (30 ml) extra-virgin olive oil

1 clove garlic, smashed

1 large head savoy cabbage, cored and thinly sliced

Kosher salt

2 cups (473 ml) low-sodium chicken broth

1 14-ounce (396 g) can cherry tomatoes

4 cups cubed (½-inch) day-old rustic Italian bread

½ cup grated Parmigiano Reggiano

Freshly ground black pepper

PAESTUM

While Pompeii may have one of Italy's most treasured archaeological sites, a little farther south lies the town of Paestum, another ancient jewel. Founded by the Greeks around 600 BC, the town was eventually taken over by the Romans. Today the area boasts a mix of temples, amphitheaters, and tombs from both cultures. Paestum offers a less traveled, slightly off-the-beaten-path glimpse into times past via almost perfectly preserved structures, exactly how they were thousands of years ago.

Grilled Eggplant Rolls with Buffalo Mozzarella

INVOLTINI DI MELANZANE GRIGLIATE CON DI BUFALA MOZZARELLA

2 small eggplant (about 8 oz./226 g each), sliced lengthwise into ⅛-inch-thick slices

3 tablespoons (44 ml) extra-virgin olive oil, plus more for the baking dish

Kosher salt and freshly ground black pepper

1 clove garlic, smashed

Pinch of crushed red pepper flakes

1 14-ounce (396 g) can cherry tomatoes

2 5-ounce (141 g) balls buffalo mozzarella, cut into ½-inch pieces

16–18 basil leaves

The Jemma family of Porta Sirena has been rearing water buffalo in the Sele plains since the early 1900s. Today their herd has grown to over 500 animals, and their cheese and meat are the staple ingredients that shape the menu at the farm's restaurant. This simple dish showcases the superior taste of mozzarella made with water buffalo milk and is well worth the splurge.

1. Prepare a medium-high gas or charcoal grill fire. Brush the eggplant slices with 2 tablespoons of the oil and season lightly with salt and pepper. Grill the eggplant, flipping once, until nicely browned and tender, 2 to 3 minutes per side. Let cool to room temperature.

2. In a 3-quart saucepan, heat the remaining 1 tablespoon of olive oil over medium-high heat. Add the garlic and cook until fragrant, about 1 minute. Add the pepper flakes, cherry tomatoes, and a pinch of salt and cook for 15 minutes. Whisk the tomato sauce to break up the tomatoes. Season to taste with salt.

3. Position a rack in the center of the oven and heat the oven to 400°F (200°C).

4. Lightly oil an 8 × 8-inch baking dish. Lay an eggplant slice on a flat surface. Top with a wedge of mozzarella and a basil leaf; roll up and transfer to the prepared baking dish, seam-side down. Repeat with the remaining eggplant slices.

5. Ladle the cherry tomato sauce over the eggplant rolls, distributing it evenly. Bake in the oven until the cheese is melted, 15 to 20 minutes. Serve immediately.

MAKES ABOUT 18 ROLLS

Water buffalo

Chocolate-Covered Almond-Stuffed Figs

FICHI RIPIENI CON MANDORLE E RICOPERTI
DI CIOCCOLATO

Fig trees thrive in southern Italy, and in the summer and fall their ripe purple and green fruits are plump and dripping with juice and hang from sagging branches, ready for picking. A common way to preserve figs is by drying them in the sun, stuffing them with almonds, and baking them. We ate them prepared in this manner at several farms. At Porta Sirena however, they take them to another level by dipping them in chocolate—delicious!

1. Position a rack in the center of the oven and heat the oven to 350°F (180°C).

2. Trim the stem from each fig. With a paring knife, cut each fig in half lengthwise, but don't cut completely through. Open the figs like a book and put one almond and a strip of lemon zest in the center of each. Press the fig together to enclose the almond and to flatten slightly.

3. Arrange the figs in a 7 × 5-inch baking dish; they should be slightly overlapping. Bake in the oven until deep golden brown, 20 to 25 minutes. Remove from the oven and let cool.

4. Put the chocolate in a bowl over a pot of barely simmering water. Once the chocolate has melted, remove from the water and let cool slightly. Dip each fig in the chocolate, turning to coat evenly. Transfer the figs to a baking sheet lined with parchment paper. Let sit at room temperature until the chocolate has set, about 2 hours. Keep stored in an airtight container.

MAKES 24

24 dried golden (Calimyrna) figs
(about 12 oz./340 g)

24 whole almonds

2 strips lemon zest, each 1 inch wide,
thinly sliced crosswise

8 ounces (226 g) bittersweet chocolate, roughly chopped

Paccheri with Sautéed Baby Artichokes and Guanciale

PACCHERI CON CARCIOFI E GUANCIALE

Paccheri are hollow tubes of pasta popular throughout Campania that resemble the bodies of calamari. They lose shape as they cook, becoming flattened, and their once hollow centers become great carriers for sauces.

1 lemon, cut in quarters

6 baby artichokes

1½ tablespoons (22 ml) extra-virgin olive oil

2 ounces (60 g) guanciale or pancetta, cut into small dice

1 clove garlic, sliced

Kosher salt

12 ounces (340 g) paccheri (or substitute rigatoni)

Freshly ground black pepper

1 tablespoon chopped flat-leaf parsley

Grated pecorino cheese, for serving

1. Squeeze the lemon quarters into a large bowl filled with cold water and then add to the bowl. Clean the baby artichokes by removing the dark green outer leaves until only the pale, tender inner leaves remain. Trim ¼ inch from the top of the artichokes, then trim the stem end and any dark parts around the bottom. Cut the artichokes in half and then into thin slices. Add to the bowl of lemon water.

2. Heat ½ tablespoon of the oil in a 12-inch skillet over medium heat. Add the guanciale (or pancetta) to the pan and cook until it has rendered its fat and is beginning to crisp, 5 to 8 minutes. With a slotted spoon, transfer to a paper-towel-lined plate. Pour off all but a thin layer of the fat.

3. Add the remaining 1 tablespoon of oil and the garlic to the pan and cook until golden brown, 1 to 2 minutes. Drain the artichokes well, and then add them to the pan with a pinch of salt. Cover, reduce the heat, and cook until the artichokes are very tender, 12 to 15 minutes. Remove the lid and add the guanciale to the pan. Set aside.

4. Meanwhile, bring a large pot of well-salted water to a boil over high heat. Drop in the pasta and cook until al dente, according to the package instructions. Reserve ¼ cup of the cooking water. Drain the pasta well and toss with the artichoke sauce over medium-low heat. Add 2 tablespoons of the reserved pasta water and toss well to combine. Add more water if the mix-

ture seems dry. Season to taste with salt and pepper. Stir in the parsley and toss well to combine. Divide among individual plates and serve with the pecorino on the side.

SERVES 4

Field of artichokes

BASILICATA

CHAPTER 9
❧ BASILICATA ❧

CARRERA DELLA REGINA

When Basilicata's fields of wheat and corn are tilled, the barren landscape impresses with its expansive openness. At times the region seems void of human existence; with abandoned farmhouses scattered about, the terrain possesses a moon-like stillness and solitude. Immersed in this unusual landscape, Carrera della Regina speaks to those seeking tranquility in an off-the-beaten-path spot.

In an area lacking the infrastructure to support tourism development, the agriturismo of the Cosentino family is an oasis of amenities and services, catering to those lucky enough to discover Basilicata's hidden charms. Impeccably restored stone barns, draped in ivy, with potted geraniums tucked into their windows, lend a refined rusticity to the property, while hearty rosemary bushes and a lush lavender garden fill the air with sweet perfume. An artificial lake excavated from a slope blends into the surroundings, and provides reprieve from the baking Basilicata sun. Lounge chairs surround its waters, and a well-stocked bar keeps guests well hydrated. Behind the lake, a long hiking trail leads deep into the woods surrounding the farm, for those seeking an immersion in nature. Dario, the youngest son of the family, manages the agriturismo's hospitality services, and emphasizes his family's goal in establishing Carrera della Regina as a premier destination for those really desiring to get away

from it all and to relax in an unspoiled environment. Chickens, ducks, and two very personable donkeys live on the grounds, as well as a large vegetable garden, but the actual working farm lies down the road. There the family raises sheep and goats, and makes fresh ricotta cheese along with an aged Toma for the farm's restaurant.

Simple recipes that rely heavily on the flavors from the land are fundamental at Carrera della Regina. The goal is to showcase a region that lies in the heart of the deep South and has historically been among Italy's poorest and most impoverished. Fresh pastas with various traditional shapes are fashioned from semolina flour and water; rich, meaty tomato sauce with veal, lamb, pork, and dried sausage is a favorite accompaniment. Generous ladlefuls are spooned over each portion, drowning the pasta in meat and tomatoes, satisfying Basilicatans' love for sauce. Dried red peppers known as cruschi are ubiquitous here; driving through the region, the peppers can be seen drying on porches and hanging from doors everywhere. They are used judiciously at the farm. One classic preparation deep-fries the peppers until they are crispy and crunchy and then crumbles them over fresh capunti pasta with shaved sheep's cheese. Potatoes, one of the vegetables that thrive in Basilicata, are boiled and tossed together with sautéed fresh peppers. They also join second courses, often lamb from the farm or barded, roasted pork loins, full of strong aromas of bay leaf and rosemary. Chicory, another important regional staple, grows wild and flourishes in the climate. It's usually paired with broad beans. In summertime cherry tomatoes are halved and top large slices of Toma cheese, then baked to produce a warm and juicy side dish. Big robust reds join the table, made from the native Aglianico grape, whose thick skin provides a tannin-laced wine that is perhaps the region's most precious and recognized culinary commodity, fetching high prices around the globe. Straightforward and nourishing, the kitchen of Carrera della Regina brings justifiable recognition to the cuisine of one of Italy's most overlooked regions. Come here to connect with a real southern Italian experience.

Basilicatan landscape

Cheese and Dried Sausage Pie

PIZZA RUSTICA

Stuffed savory pies are common throughout southern Italy, especially around Easter time. In this version, the chewy, yeasted dough is a nice counterpoint to the soft cheese filling. It is best served warm or at room temperature—although at home, we have been known to sneak pieces cold from the refrigerator for a very non-Italian breakfast treat.

1. **Make the dough:** Combine the flour and salt in a large bowl. Make a well in the center of the flour and add the milk, oil, and yeast. Lightly beat the liquid ingredients with a fork to combine. Gradually pull in some of the flour mixture until a dough begins to form. Turn the dough out onto a clean surface and knead until smooth and elastic, 3 to 5 minutes. Divide the dough in half. Roll out one half to a 12-inch round and place it inside a 9-inch pie plate.

2. **Make the filling:** In a large bowl, combine the Toma cheese, ricotta, dried sausage, eggs, Parmigiano, and salt.

3. **To assemble,** spread the cheese filling evenly over the pie dough.

4. Roll out the other half of dough to a 10-inch circle and place it over the filling, trimming away any excess. Crimp together the edges of dough, and then pierce all over with a fork.

5. Position a rack in the center of the oven and heat the oven to 350°F (180°C).

6. Let the pie sit at room temperature for 30 minutes. Beat the egg yolk with 1 teaspoon of water and brush over top crust. Bake the pie in the oven until deeply golden and the crust is set, 30 to 35 minutes. Transfer to a rack and let cool for 10 minutes before slicing and serving.

SERVES 6–8

For the dough:

2 3/4 cups (344 g) all-purpose flour

1 teaspoon Kosher salt

1 cup (240 ml) warm milk

3 1/2 tablespoons (51 ml) extra-virgin olive oil

1 teaspoon active dry yeast

For the filling:

6 ounces (170 g) Toma cheese, cut into 1/2-inch dice (about 3/4 cup)

1 heaping cup (250 g) ricotta cheese

4 1/2 ounces (130 g) dried sweet Italian sausage, quartered lengthwise and then sliced thinly (about 1 cup)

2 eggs, lightly beaten

1/2 cup freshly grated Parmigiano Reggiano cheese

1 teaspoon Kosher salt

To assemble:

1 egg yolk

Capunti Pasta with Dried Sweet Italian Peppers

CAPUNTI CON CRUSCHI

1 recipe semolina pasta dough (see "Basic Recipes")

6 tablespoons (88 ml) extra-virgin olive oil

2 cloves garlic, sliced

2 pints grape tomatoes, halved if large

2 tablespoons chopped flat-leaf parsley

Kosher salt and freshly ground black pepper

4 dried sweet peppers, such as cruschi

3 ounces (85 g) ricotta salata, shaved

1. Divide the dough into eight pieces and roll out each piece into a ¼-inch-thick rope. With a pastry cutter or knife, cut the dough into 1-inch pieces. With your three center fingers, press down on a piece of dough and then drag the dough across the work surface toward your body to make an indentation. Transfer to a rimmed baking sheet lightly dusted with flour, and continue with the remaining pieces of dough. Set aside until needed.

2. Heat 2 tablespoons of the oil in a straight-sided 11-inch skillet over medium heat. Add the garlic and cook until fragrant and lightly golden, about 2 minutes. Add the tomatoes and cook, stirring occasionally, until they begin to break down, 12 to 15 minutes. Toss in the parsley and season with a generous pinch of salt and a few grindings of pepper.

3. In a 10-inch skillet, heat the remaining ¼ cup of oil over medium heat. Add the dried peppers and cook until fragrant and crisp, 1 to 2 minutes. Transfer to a paper-towel-lined plate.

4. Bring a large pot of well-salted water to a boil. Drop in the pasta and cook until they begin to rise to the surface and are tender, 6 to 8 minutes. Reserve ¼ cup of pasta water, then drain the pasta well.

5. Toss the pasta with the tomato sauce, adding the reserved pasta water as needed if the sauce seems dry. Season to taste with salt and pepper. Crumble the peppers over the pasta and then top with the shaved ricotta. Serve immediately.

SERVES 4

Capunti Pasta with Cruschi Peppers

CRUSCHI

In early fall the doors and balconies of houses throughout Basilicata are decorated with strung red peppers that dangle and shrivel in the hot southern sun. Peppers thrive here during the summer months and are sautéed with garlic and olive oil and served over boiled potatoes for a common side dish. The vegetables are preserved for the winter months by hanging to dry; once dried, they're referred to as cruschi. A classic preparation involves frying the dried peppers with a sprinkle of salt to produce a crispy, crackly, sweet, and smoky treat that can be served as an antipasto with a sharp cheese or crumbled and sprinkled over pasta or meat. The burnt red capsicum has become Basilicata's most emblematic ingredient, eaten until fresh peppers grow again in the warmer months. Their addictive taste and crunchy texture rivals any potato chip on the market—and the peppers are nearly as much fun to say (*croooski*) as they are to eat.

Drying cruschi peppers

Pork Bracciole with Eggplant and Scamorza Cheese

BRACCIOLE DI MAIALE CON MELANZANE E SCAMORZA

We had never thought to add a vegetable to bracciole, but grilled eggplant's presence with the rolled-up meat really makes this dish great.

1. Prepare a medium-high gas or charcoal grill.

2. In a large bowl, toss the eggplant with 2 tablespoons of the oil and season with salt and pepper. Grill the eggplant until browned on both sides and tender, 6 to 8 minutes total. Let cool completely.

3. Arrange the pork cutlets on a cutting board and season generously with salt and pepper. Place two slices of eggplant on each pork cutlet and then top with scamorza cheese. Fold the sides of the pork in toward the center, then roll up the pork lengthwise. Tie each bundle with kitchen string. Set aside on a plate.

4. Crush the tomatoes with your hands in a large bowl and set aside. Heat the remaining tablespoon of oil in a 12-inch straight-sided pan over medium heat. Add the bracciole, in batches, and brown all over, 2 to 3 minutes per side. Transfer to a large plate. Add the garlic and cook until golden brown all over, 2 to 3 minutes. Add the tomatoes and bring to a boil. Add the pork, any accumulated juices, and the bay leaf and return to a boil. Reduce the heat to maintain a gentle simmer and cook until the pork is tender, about 1 hour.

5. Remove the pork from the pan and place on a serving platter, removing the kitchen string with scissors. Season the sauce to taste with salt and pepper, then ladle the sauce over the pork and serve.

SERVES 6

1 medium eggplant, sliced into 1/8-inch rounds

3 tablespoons (44 ml) extra-virgin olive oil

Kosher salt and freshly ground black pepper

12 pork cutlets (2 1/4 – 2 1/2 pounds/ 1–1 1/4 kg total), pounded to 1/8 inch thick

6 ounces (170 g) scamorza cheese, shredded

1 28-ounce (793 g) can whole plum tomatoes

1 clove garlic, smashed

1 fresh bay leaf

Poached Pear and Ricotta Mousse Tart

TORTA DI PERE E RICOTTA

1⅓ cups (150 g) all-purpose flour

⅓ cup toasted slivered almonds

Pinch of table salt

½ cup (1 stick) unsalted butter, softened

⅓ cup superfine sugar

1 large egg

1 cup granulated sugar

2 3-inch strips orange zest

1 small cinnamon stick

3 small firm-ripe Forelle pears (about 1 pound/453 g total), peeled, halved, and cored

1 cup (240 ml) heavy cream

8 ounces (226 g) ricotta cheese, drained

2 tablespoons chopped unsalted pistachio nuts

This tart's rich buttery crust is the perfect carrier for a smooth ricotta mousse. The poached pears add sweetness and round out the flavors, creating a satisfying and delicious treat.

1. Butter a 4 × 13½-inch rectangular fluted tart pan with a removable bottom.

2. Combine the flour, almonds, and salt in a food processor and process until the almonds are finely chopped. In a stand mixer fitted with the paddle attachment, cream together the butter and superfine sugar at medium speed until light and fluffy, 2 to 3 minutes. Add the egg and beat another 45 seconds. Reduce the speed to low and add the flour mixture; beat until a dough forms. With lightly floured hands, press the dough evenly over the bottom and up the sides of the prepared tart pan. Prick the dough all over with a fork and then refrigerate until firm, about 1 hour.

3. Position a rack in the center of the oven and heat the oven to 375°F (190°C).

4. Line the dough with parchment paper and fill with pie weights or dried beans. Bake until the edges are just beginning to turn golden brown, 10 to 12 minutes. Remove the weights or beans and the parchment paper and continue baking until the crust is deeply golden and the center of the crust is dry, 5 to 7 minutes. Transfer to a rack and let cool completely.

5. Meanwhile, combine 3 cups of water with the granulated sugar, orange zest, and cinnamon in a medium saucepan. Bring to a boil over medium-high heat, stirring often, until the sugar melts. Add the pears and reduce the heat to maintain a gentle simmer; cook until the pears are tender when pierced with a skewer, 15 to 20 minutes. Remove from the heat and let the pears cool completely in the syrup. With a slotted spoon,

remove the pears from the syrup and cut lengthwise into ¼-inch-thick slices. Discard the orange zest and cinnamon stick. Return the saucepan to medium-high heat and boil until the liquid is reduced by half, 5 to 8 minutes. Set the pan aside.

6. In a large chilled bowl, beat the heavy cream with a handheld mixer until stiff peaks form. Set aside. In the bowl of a stand mixer fitted with a paddle attachment, beat the ricotta cheese with 2 tablespoons of the poaching syrup on medium speed until thick and creamy. Fold the whipped cream into the ricotta mixture with a spatula. Spread the mixture out onto the baked tart shell, arrange the sliced pears over the top, and brush with some of the remaining poaching liquid. Refrigerate the tart for at least 2 hours. Sprinkle with the nuts before serving.

SERVES 8

Poached Pear and Ricotta Mousse Tart

CALABRIA

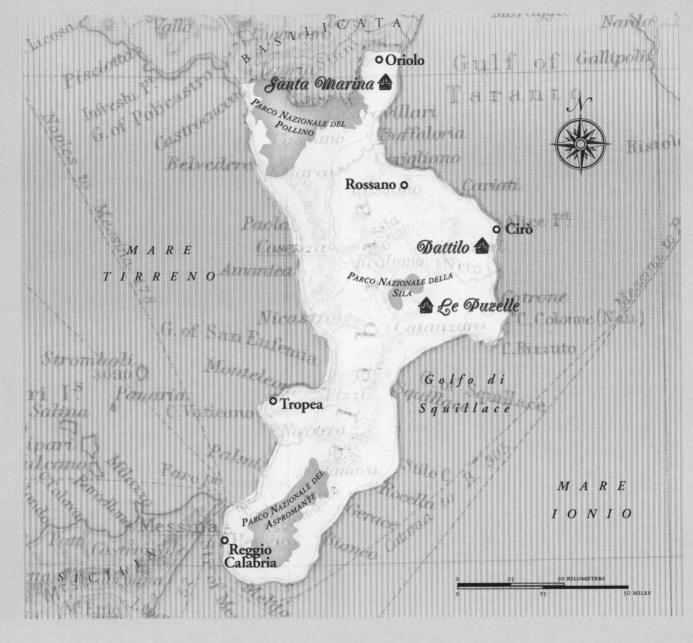

CHAPTER 10
CALABRIA

SANTA MARINA

Straddling the border of Basilicata and surrounded by the mountains and valleys of the Pollino, Italy's largest national park, the Santa Marina agriturismo offers a getaway into the remote, often overlooked northeast corner of Calabria. The farm's allure lies within its spectacular setting: a stunning backdrop of soft hills strewn with sheep and stone farmhouses, valleys cut by flowing streams, tiny hamlets, winding roads often blocked with cattle, and distant views of the shimmering Ionian Sea. Rustic amenities and the genial Pina, Santa Marina's owner, greet guests who visit the farm for its solace and serenity.

Shortly after their marriage, Pina and her husband, Salvatore, left the struggling economy of Calabria and migrated north to Milan, where there was opportunity for work. There they found jobs and raised a family, never forgetting their roots and hoping for the ability to one day return to the south. Eventually they followed their hearts and came back to Calabria to help out on Pina's family farm. Upon returning, they found a different Calabria than the one they had left. Many of the neighboring farms neglected and vacant, with younger generations leaving the countryside for more prosperous lives in the cities. To help supplement their income, the couple decided to open their doors to tourism. They

restored the barns into individual bedrooms and the basement of the main farmhouse into a large restaurant, and in 1994 they opened the Santa Marina agriturismo. To Pina, this has been a dream come true. With unbridled energy and passion, she handles the never-ending daily chores of running her farm with a smile on her face. From feeding her chickens, rabbits, and pigs, to tending her vegetable gardens, fruit trees, and olive groves, to tidying the guest bedrooms and cooking dinner, she maintains a frenetic pace that's difficult to keep up with.

Pina's cooking respects customary dishes from both Calabria and neighboring Basilicata, yet is inventive, resourceful, and centered on ingredients produced on her own farm. Fallen fruit from apple and pear trees, collected acorns from surrounding oak trees, and even leftover pasta water are all fed to Santa Marina's pigs, which creates rich flavorful meat used for making cured salami, prosciutto, and pancetta, while wild greens are given to the rabbits for a more natural diet. Veal and cheese are sourced from an attached farm owned by an elderly couple who continue to till their land with oxen. A large vegetable garden provides the southern Italian staples of eggplant, zucchini, and tomatoes, as well as an abundance of the essential Calabrian hot peppers. Pina cooks the fiery capsicums with sugar to make an intensely spicy jelly that accompanies dishes, and plates

of the chilies accompany steaming bowls of pasta fagioli for anyone wishing to add a little heat to the dish. She fries zucchini flowers, stuffs eggplant and tomatoes, and sautés peppers with eggs and pancetta to satiate the Calabrian hunger for fresh vegetables. Pasta is made daily by hand, with only water and semolina flour for the dough, symbolic of the region's impoverished past, when eggs were considered too much of a delicacy for everyday use. Sauces are predominantly vegetable-based, whether fresh in summer or jarred during the winter. Dessert usually includes fruit from the numerous pomegranate, fig, and persimmon trees that thrive in the mild climate around Santa Marina and is followed with Pina's own secret recipe for an after-dinner liquor made from olives. The humble, potent shot serves as a metaphor for this isolated corner of Calabria, where simple ingredients are stretched to create a bucolic cuisine full of honest flavor and goodness.

Cured Calabrian pork products

Eggplant and Zucchini Rolls
INVOLTINI DI MELANZANE E ZUCCHINE

A mandoline will help you slice the vegetables into even, thin strips for this tasty summer appetizer.

1. Prepare a medium-high gas grill. Brush the zucchini and eggplant slices with 2 tablespoons of the oil and season with 1 teaspoon salt and a few grinds of pepper. Grill the zucchini and eggplant on both sides until tender and nicely browned, about 3 minutes per side. Transfer to a large plate.

2. Meanwhile, in a medium bowl, combine the Parmigiano, bread crumbs, basil, parsley, capers, nutmeg, and oregano. Stir in the chopped tomatoes and the remaining 1 tablespoon of oil. Season to taste with salt and pepper.

3. Place 1 strip of zucchini vertically on a cutting board. Add 1 tablespoon of filling at the bottom of the strip, leaving a margin of about ¼ inch around the edges. Roll the zucchini up and fasten with a toothpick. Continue with the remaining zucchini and eggplant in the same manner. Arrange the rolls on a serving platter and drizzle with a little oil.

SERVES 4

1 medium zucchini, sliced lengthwise ¼ inch thick

1 medium eggplant, sliced lengthwise ¼ inch thick

3 tablespoons (44 ml) extra-virgin olive oil, plus more for serving

Kosher salt and freshly ground black pepper

½ cup finely grated Parmigiano Reggiano

¼ cup bread crumbs

3 tablespoons finely chopped basil

2 tablespoons finely chopped flat-leaf parsley

2 tablespoons finely chopped capers

¼ teaspoon freshly grated nutmeg

Pinch of dried oregano

6 grape tomatoes, finely chopped

Rascatelli with Spicy Tomato Sauce

RASCATELLI AL POMODORO PICCANTE

1 recipe semolina pasta dough (see "Basic Recipes")

2 tablespoons (30 ml) extra-virgin olive oil

1½ pounds (680 g) cherry tomatoes, quartered

1 small fresh chile pepper, thinly sliced

2 tablespoons thinly sliced basil

Kosher salt

Freshly grated cacioricotta or ricotta salata cheese, for serving

This pasta shape is traditional of the Basilicata region and illustrates the melding of regions in Pina's cooking. The pasta takes its name, rascatelli, from the scratchy sound your fingers make against the wooden board when you drag each piece of dough across to make its signature indentation.

1. Divide the dough into eight pieces, then roll out each piece into a rope ½ inch thick. With a dough scraper or knife, cut the dough into ½-inch pieces. Use your middle finger to press down on the center of each piece and drag the dough toward your body, making an indentation. Repeat with the remaining pieces of dough, transferring the pasta to baking sheets lined with parchment paper.

2. Meanwhile, make the sauce: Heat the oil in a 12-inch skillet over medium-high heat. Add the tomatoes, chile pepper, and basil; cook until the tomatoes are just starting to break down, 6 to 8 minutes. Season with a generous pinch of salt and continue cooking until the tomatoes are completely broken down. Taste for seasoning.

3. Bring a large pot of well-salted water to a boil. Drop the pasta into the water and cook until tender, 10 to 13 minutes. Reserve ¼ cup of the pasta water. Drain the pasta well and then toss immediately in the sauce, adding the reserved water as needed. Sprinkle with the freshly grated cheese and serve.

SERVES 6

POLLINO PARK

Santa Marina is set amid isolated, protected beauty in the confines of Italy's largest green space, Pollino National Park. Straddling both Basilicata and Calabria, the territory is immense and pocketed with dense forests and mountain peaks. The villages that lie in the shadow of Monte Pollino, the park's highest peak at 7,382 feet, have a northern alpine feeling to them and are lined with pine trees. Stacks of wood are piled outside houses, and the air is redolent with smoke. Cows and horses graze freely in lush fields. The park is also home to a rare pine tree originating in the Balkans known as pino loricato. In Italy these trees grow solely in the park and stand as its symbol, with their gnarled scaly trunks dotting the snow-peaked mountains. A day spent hiking here, absorbing the splendid scenery, awakens the senses and stirs the appetite for dinner at Pina's, whose cooking suits the austere landscape.

Sausage Cooked in Red Wine with Sweet Dried Peppers and Black Olives

SALSICCE AL VINO ROSSO CON PEPERONI
SECCHI E OLIVE

¼ cup (59 ml) extra-virgin olive oil

4 sweet Italian sausage links (about 4 oz./115 g each), pricked all over with a fork

½ cup (118 ml) dry red wine

6 dried sweet peppers, stemmed, seeded, and cut into ¾-inch-thick slices

Kosher salt

⅓ cup small black olives, such as niçoise or Gaeta

This dish showcases the Basilicatan influences on Pina's cooking with its use of the region's prized sweet dried peppers known as cruschi. We have found that any dried nonspicy pepper works well to replicate this dish—but be careful not to burn the vegetable's delicate skin in overheated oil.

1. Heat 1 tablespoon of the oil in a 12-inch skillet over medium heat. Add the sausages and brown all over, about 5 to 7 minutes. Add the red wine, cover, reduce the heat to medium-low, and continue cooking until the sausages are cooked through and the wine is thick and syrupy, about 7 to 10 minutes. Transfer the sausages with the wine sauce to a platter, and then wipe out the skillet.

2. Heat the remaining 3 tablespoons of oil in the skillet over medium heat until shimmering. Add the peppers and a pinch of salt; cook, stirring continuously, until the peppers are lightly toasted, about 1 minute. With a slotted spoon, transfer the peppers to a clean plate. Add the olives to the oil and cook until warmed through, about 1 minute. Using the slotted spoon transfer the olives to the plate with the peppers.

3. Cut the sausages into slices on a bias and place on a platter. Top with the peppers and the olives. Drizzle with the wine sauce and serve.

SERVES 4

Boiled Potatoes with Garlicky Peppers

PATATE ALL'AGLIO E PEPERONI

This recipe's utter simplicity showcases the Italian knack for taking a few ingredients and turning them into a wholesome and satisfying dish.

1. Put the potatoes and ½ tablespoon salt in a 4-quart saucepan and bring to a boil over medium-high heat. Reduce the heat and simmer until the potatoes are tender when pierced with a fork or skewer, 20 to 30 minutes. Drain the potatoes; when they're cool enough to handle, peel them and then slice them into rounds. Transfer the potatoes to a large serving bowl.

2. Heat the oil in a 12-inch skillet over medium-high heat until shimmering. Add the garlic and cook until fragrant, about 1 minute. Add the peppers and a ¼ teaspoon salt and cook until the skin blisters and the peppers are tender, 2 to 4 minutes. With a slotted spoon, transfer the peppers to the bowl with the potatoes. Drizzle the oil over the top, sprinkle with a generous pinch of salt, and serve.

SERVES 4

2 medium Yukon Gold potatoes

Kosher salt

3 tablespoons (44 ml) extra-virgin olive oil

1 clove garlic, sliced

6 Italian long hot peppers, cut into ½-inch slices

Prosciutto and Caciocavallo Stuffed Veal Rib Chops with Shiitake Sauce

COSTOLETTE DI VITELLO RIPIENE

4 bone-in rib veal chops (about
  12 oz./340 g each), 1 inch thick

Kosher salt and freshly ground black
  pepper

4 thin slices prosciutto

4 thin slices caciocavallo cheese

¼ cup (59 ml) extra-virgin olive oil

2 cloves garlic, smashed

1 pound (453 g) shiitake mushrooms,
  stemmed and sliced into ¼-inch
  strips

½ cup (118 ml) white wine

1 tablespoon chopped rosemary

1 tablespoon chopped thyme

1. Position a rack in the center of the oven and heat the oven to 425°F (220°C).

2. With a sharp knife, make a horizontal slit in each veal chop to create a 2-inch-long pocket. Season the chops generously all over with salt and pepper, including the inside of the pocket. Stuff each chop with 1 slice of prosciutto and 1 slice of caciocavallo cheese. Secure the pockets with a toothpick.

3. Heat 2 tablespoons of the oil in a 12-inch skillet over medium-high heat until shimmering. Sear the chops on both sides until deeply golden brown, 3 to 5 minutes per side. Transfer the chops to a baking sheet and roast in the oven until an instant-read thermometer inserted into the thickest part of the veal registers 140°F (60°C), 10 to 12 minutes. Remove from the oven and tent loosely with foil.

4. Meanwhile, heat the remaining 2 tablespoons of oil to the skillet used to cook the veal over medium-high heat. Add the garlic and cook until golden brown, 1 to 2 minutes. Discard the garlic. Add the mushrooms and a ½ teaspoon of salt and sauté until they begin to soften, 5 to 7 minutes. Add the white wine, scrape up any browned bits from the pan with a wooden spoon, and reduce until the liquid is almost gone, 2 to 3 minutes. Stir in the rosemary, thyme, and ½ cup of water and cook until reduced by half. Pour in any juices that have accumulated from the veal chops and season to taste with salt and pepper.

5. Put the chops on individual plates and top with the mushroom sauce. Serve immediately.

SERVES 4

DATTILO

During the fall harvest at the Roberto Ceraudo winery and agriturismo, the intoxicating aroma of just-pressed grapes permeates the compound of ancient stone houses. The air is rich and heady, testimony to the natural process under way in the cantina, where vats of grape juice quietly ferment under the watchful eye of owner Roberto Ceraudo. With a scruffy beard and a gently wrinkled, yet youthful face, Roberto wears a seasoned look that speaks of his nearly forty years of winemaking experience. A legend and visionary among Italian winemakers, Roberto, a steadfast purist, set out to distinguish himself from others in Calabria by adopting organic practices from the beginning. This risky endeavor has proved its worth, as the farm's healthy fruit and delicious wines have received countless accolades and awards since the winery's inception in 1973. Production ranges from a line of indigenous varietals—including big tannic reds from the Gaglioppo grape—to citrus and acidic whites pressed from local Mantonico grapes, to international blends of Cabernet Sauvignon and Chardonnay. The ancient practice of planting rosebushes in each row of vines is still used in the vineyard today; the flowers warn growers if there are parasites since they attack the roses first. No chemicals or pesticides treat the vines. This all-natural approach, coupled with a microclimate of hot sunny days and cool nights generated from the nearby Ionian Sea, produces fruit that needs little manipulation in the cellar. Roberto has also learned many winemaking secrets over the years. One golden rule of the winery, for instance, is that during the September harvest, work begins before dawn in the cool morning, before the scorching Calabrian sun rises and threatens to begin fermenting the grapes before they are pressed. A similar organic approach has been applied to the farm's olive trees, which have earned the Roberto Ceraudo farm top prize in numerous tastings for the oil's near zero percent acidity. Roberto's artisanal approach to what he does has been passed along to his children, whose passion for the family business ensures the future. Working side by side with his father, Giuseppe divides his time between the vineyard and the cantina, while eldest daughter Susy helps with marketing and sales. The youngest daughter, Caterina, is finishing a degree in oenology and hopes to one day oversee all winemaking.

A desire to bring their wines and olive oil together in a graceful setting prompted the family to found the Dattilo agriturismo and restaurant. In an open, white stone room, guests dine in refined elegance beneath dark wooden beams and stone arches. White linen tablecloths, dim lighting offset by flickering candles, a fire stoked in the original stone fireplace, and a grand piano in the center of the room lend a luxurious New York City feel to Dattilo, which seems worlds away from a working farm in Calabria. Yet directly outside the dining room door lie vegetable gardens, fruit trees, acres of vineyards, groves of olive trees, and free range chickens. This is a genuine farm restaurant.

Dattilo was a pioneer in Calabria, one of the first restaurants to offer avant-garde cuisine in a region known for its peasant cooking traditions. Three different levels of tast-

ing menus are on offer, paired with a wine list representing all of Italy. Stylish glass and ceramic plates suit the contemporary crafted cuisine and out of the kitchen emerges innovative combinations of Calabrian ingredients. After a twenty-four-hour warm-water sous-vide bath, lamb blade chops are perched atop a pecorino cheese puree infused with licorice from a nearby manufacturer. Humble salt cod becomes something elaborate when dressed in a light crispy tempura batter and set against a backdrop of creamy beans, sweet candied lemon peel, and a dusting of coffee to add a hint of bitterness. A local spicy salami, 'nduja, gets shaped into rolls that come served warm, fresh from the oven with a cold emulsification of the farm's olive oil that turns to liquid when spread over the steaming bread.

Sardella, a spicy fish sauce made from newly hatched sardines, is drizzled over crispy phyllo packets of tomatoes and provolone that melt in your mouth, leaving a lingering salty taste of the Ionian Sea.

Yet aside from the elevated level of its wine and food, perhaps what makes Dattilo such a special place is the jovial nature of everyone who works there. With all of the recognition Roberto has received for his life's achievements, his humility and endearing manner are infectious. Everyone at Dattilo, from his children to the farmers, office staff, waiters, and cooks, shares this hospitable spirit. When you see Roberto making the rounds of his restaurant, joking and bringing life to the room, his passion for what he does is evident.

Crispy Phyllo Packets with Cherry Tomatoes and Provolone

SFOGLIATELLE CROCCANTI CON POMODORINI E PROVOLA

Flaky phyllo dough is the perfect match to a gooey center made from provolone and tomatoes in this juicy dish. At Dattilo, they serve these crispy packages drizzled with a local fish sauce known as sardella to add a salty note.

1. Position a rack in the center of the oven and heat the oven to 400°F (200°C).

2. In a medium bowl, mix together the cherry tomatoes, provolone, thyme, ½ teaspoon of salt, and a few grinds of pepper.

3. Melt the butter in a small skillet over medium heat.

4. On a clean work surface, lay out 1 sheet of phyllo dough and brush the sheet all over with the melted butter. Top with another sheet of dough, again brushing with butter; continue layering the phyllo sheets in this manner, ending with the fourth sheet. Cut the dough into 16 rectangles each 4 × 2⅞ inches. Place 1 heaping tablespoon of filling in the center of each rectangle. Fold one of the longer ends of the rectangle up over the center of the filling, and then fold the other long end up to meet it. Fold the two smaller ends under the packet. Brush the top with more melted butter and sprinkle with sesame seeds. Transfer to a baking sheet lined with parchment, and continue with the remaining packets.

5. Bake the packets in the oven until they are a deep golden brown, 10 to 13 minutes. Remove from the oven and let cool on a rack for 5 minutes before serving.

MAKES 16 PACKETS

1 cup (about 7 oz./200 g) cherry tomatoes, cut into ⅛-inch pieces

½ cup (3 oz./90 g) provolone cheese, cut into ⅛-inch dice

1 teaspoon chopped fresh thyme

Kosher salt and freshly ground white pepper

2 tablespoons (¼ stick) unsalted butter

4 sheets phyllo dough

1 tablespoon toasted sesame seeds

Phyllo packets

Grilled Calamari with Fava Bean Puree and Sautéed Chicory

CALAMARI ALLA GRIGLIA CON CICORIA E CREMA DI FAVE

The Ionian Sea lies a few miles away from Dattilo, providing the kitchen with a bounty of fresh local seafood. Grilling calamari adds a pleasant charred flavor that suits the bitter greens and the strong flavor of fava bean puree in this recipe.

4 ounces (115 g) dried fava beans, soaked overnight

1 medium yellow onion, halved

1 carrot, peeled

1 stalk celery

2 fresh bay leaves

Kosher salt

¼ cup (59 ml) extra-virgin olive oil, plus more for serving

Freshly ground black pepper

1 large bunch (about 1 pound/ 453 g) chicory, trimmed and thinly sliced

1 clove garlic, smashed

Pinch of hot pepper flakes

8 whole calamari, cleaned

1 lemon, cut into wedges

1. Drain the beans, peel the skins, and rinse well. In a 6-quart pot, add the favas, onion, carrot, celery, and bay leaves; cover with water by 1 inch (about 2 quarts of water). Bring to a boil over medium-high heat, reduce the heat to maintain a simmer, and cook until the fava beans are tender, 35 to 40 minutes. Reserving ¼ cup of the cooking liquid, drain the fava beans; discard the vegetables and bay leaves. Puree the fava beans with 2 tablespoons of the oil and the reserved broth until smooth. Season to taste with salt and pepper.

2. Bring 4 quarts of well-salted water to a boil. Add the chicory and cook until just tender, about 2 minutes. Drain well, and then immediately submerge into a bowl of ice water. Drain again and dry on kitchen towels. Heat 1 tablespoon of olive oil in a 10-inch skillet over medium heat. Add the garlic and cook until golden brown, 3 to 4 minutes. Discard the garlic. Add the pepper flakes to the pan and heat until fragrant, about 30 seconds. Add the chicory and cook until heated though and coated with the oil, 2 to 3 minutes. Season to taste with salt.

3. Prepare a medium-hot gas grill or a grill pan over medium heat. In a large bowl, toss the calamari with the remaining 1 tablespoon oil, 1 teaspoon salt, and a few grinds of pepper. Grill the calamari until grill marks form, about 2 minutes. Flip and cook until grill marks form and the calamari is cooked through, 2 to 3 minutes.

4. To serve, divide the chicory among four plates, top with a spoonful of fava bean puree, then top each plate with two calamari. Drizzle with olive oil and serve with the wedges of lemon.

SERVES 4

Drying grapes for dessert wine

Shrimp-Stuffed Zucchini Flowers

FIORI DI ZUCCA RIPIENI CON GAMBERI

8 ounces (226 g) large shrimp, preferably wild, peeled and deveined

12 grape tomatoes, finely chopped

1 teaspoon chopped thyme

2 tablespoons (30 ml) extra-virgin olive oil

Kosher salt and freshly ground black pepper

12 zucchini flowers, washed and trimmed

Margherita, the chef at Dattilo, boils, blanches, and purees the tender inner leaves of the zucchini plant to act as a verdant green bed for these oven-roasted shrimp-stuffed flowers. We found that they are just as good without the sauce and like them with a squeeze of lemon juice, a drizzling of good olive oil, and a sprinkling of salt.

1. Position a rack in the center of the oven and heat the oven to 350°F (180°C).

2. Finely chop the shrimp and then transfer to a medium bowl. Add the tomatoes, thyme, 1 tablespoon of the oil, and a pinch of salt and pepper; mix well. Fill the zucchini flowers with 1 heaping tablespoon of the shrimp filling. Arrange on a small baking dish and then drizzle with the remaining oil and sprinkle with a pinch of salt.

3. Bake in the oven until the flowers are lightly golden, and the shrimp is cooked through, 7 to 10 minutes. Remove from the oven.

MAKES 12 STUFFED FLOWERS

Stuffed Zucchini Flowers

Checking grapes to harvest

Shrimp Risotto with Reduced Citrus Sauce

RISOTTO AI GAMBERI CON SALSA DI AGRUMI

1½ pounds (680 g) large shrimp
with shells

3 tablespoons (44 ml) extra-virgin
olive oil

1 small red onion, thinly sliced

1 carrot, peeled and sliced

Kosher salt

1 cup cherry tomatoes

1¼ (300 ml) cups dry white wine

2 sprigs flat-leaf parsley

2 thyme sprigs

1 fresh bay leaf

½ cup (118 ml) freshly squeezed
orange juice, strained

1 tablespoon (15 ml) lemon juice

2 tablespoons (¼ stick) unsalted
butter

1 medium yellow onion, cut into
fine dice

2 cups Arborio rice

Chopped flat-leaf parsley, for
garnish

While risotto may seem like a more northern Italian dish, Calabria actually cultivates its own rice in an area known as the plains of Sibari. Fifteen varieties of rice are grown, and the grain has become a somewhat traditional product to the region. At Dattilo, they make an intensely flavored shrimp risotto with Arborio rice from the Sibari. A slightly concentrated shrimp stock adds depth to this creamy first course, and the citrus reduction adds brightness and acidity.

1. Peel and devein the shrimp. Reserve the peels and set aside. Cut the shrimp into ¼-inch pieces and refrigerate until needed.

2. Heat 1½ tablespoons of the olive oil in a 6-quart pot over medium-high heat. Add the onion, carrot, and pinch of salt, and sauté until just tender, 5 to 6 minutes. Add the cherry tomatoes and shrimp shells and cook until lightly browned, 4 to 5 minutes. Add ¾ cup of the white wine and reduce by half, 2 to 3 minutes. Add 1½ quarts of water along with the parsley, thyme, and bay leaf. Bring to a boil, then reduce the heat and gently simmer, skimming any scum that rises to the surface, until the broth is fragrant, 35 to 40 minutes. Remove from the heat and set aside to cool slightly. Strain the broth through a fine-mesh sieve into a 4-quart pot, pressing down on the solids to extract as much flavor as possible. Discard the solids and keep the broth warm.

3. Put the orange juice, lemon juice, and a pinch of salt into a small saucepan. Bring to a boil over medium-high heat and reduce by half, 5 to 8 minutes. Swirl in the butter and simmer until the sauce is thick and shiny, 2 to 3 minutes.

4. Heat the remaining 1½ tablespoons of oil in a heavy 12-inch straight-sided sauté pan over medium heat. Add the yellow onion and a pinch of salt and sauté until the onion is tender and translucent but not browned, 8 to 10 minutes. Add the rice and stir to coat the rice with the oil. Cook the rice until opaque, 2 to 3 minutes. Add the remaining ½ cup of wine, and reduce until dry.

5. Add a ladle of the shrimp stock to cover the rice, and stir continuously until the stock has reduced to below the rice. Add another ladle of stock to cover, and continue cooking the rice in this manner for 20 minutes, stirring continuously. Add the shrimp to the rice and continue to cook, adding stock one ladle at a time, until the shrimp is cooked through, 3 to 5 minutes more.

6. Divide the risotto among individual shallow bowls. Drizzle each plate with the citrus sauce and garnish with chopped parsley.

SERVES 4–6

Chef Margarita making shrimp bisque

Citrus and Caramelized Almond Semifreddo
SEMIFREDDO AGLI AGRUMI E MANDORLE CARAMELLATE

Drying orange and lemon zest helps soften the bitterness and concentrates the intense flavors. This cool and creamy egg-rich semifreddo is the perfect match to the bright notes of the citrus—and it gets even better when topped with caramelized almond brittle.

4 4-inch strips lemon zest, white
   pith removed

4 4-inch strips orange zest, white
   pith removed

½ cup slivered almonds

1 cup plus 2 tablespoons granulated
   sugar

10 large egg yolks

1 ½ cups (360 ml) heavy cream

1. Position the rack in the center of the oven and heat the oven to 350°F (180°C).

2. Arrange the strips of lemon and orange zest on a baking sheet. Place in the oven to dry out, 13 to 15 minutes. Remove from the oven and let cool completely. Finely chop up the zest and mix together in a small bowl.

3. Put the almonds on a baking sheet and toast in the oven until lightly golden, about 5 minutes. Remove from the oven and let cool.

4. In a small saucepan, melt ¼ cup of the sugar over medium-high heat without stirring. Cook the sugar until it turns amber, 5 to 6 minutes. Add the almonds to the pan and stir to coat. Pour the mixture out onto a piece of parchment paper. With an offset spatula, spread the mixture out into an even layer. Let the mixture cool slightly and then break it into small pieces. Put the almond brittle in a food processor and pulse until finely chopped. Transfer to a bowl and set aside.

5. In the pan of a double boiler, combine the egg yolks and ¾ cup of sugar. Place the pan over barely simmering water and whisk constantly until the mixture is thickened and doubled in volume and its temperature reaches 175°F (80°C), about 4 minutes. Remove the pan from the heat and immediately submerge in an ice-water bath, stirring constantly to bring down the temperature.

6. Meanwhile, in a stand mixer fitted with the whisk attachment, beat the heavy cream on medium speed until soft peaks form. Gradually add the remaining 2 tablespoons of sugar in a steady stream, and continue to beat until stiff peaks form. Transfer to a large bowl.

7. In a stand mixer fitted with the paddle attachment, beat the egg yolk mixture with the citrus zest until thickened and very pale, 3 to 5 minutes.

8. Using a spatula, add a quarter of the whipped cream to the egg mixture and stir together gently to lighten the base. Fold the remaining whipped cream into the egg mixture and then spoon the custard into eight 4-ounce ramekins. Cover with plastic wrap and freeze until firm, about 4 hours.

9. To serve, remove the ramekins from the freezer and let stand at room temperature for 3 minutes. Run a butter knife around the edge of each ramekin, then invert each semifreddo onto an individual plate. Sprinkle each top generously with the ground almond brittle and serve.

AMARELLI LICORICE

The eastern coast of Calabria harbors a specific climate perfectly suited for growing the delicate plant that is the base of a seemingly very non-Italian delicacy. Licorice has long been a favorite treat among Calabrians, who have been producing candy from the root's extracted juices since the sixteenth century. The Amarelli family of Rossano has been transforming the plant into candy for over 200 years; their name has become synonymous with authentic high-quality licorice. Their factory and museum showcase the process involved in producing the root. A Calabrian specialty incorporates the juices into an intensely concentrated after-dinner liquor known as Liquirizia di Calabria, which is poured into espresso to offer a caffeinated herbal jolt. Modern chefs have also been turning to licorice to add anise flavor to their cooking. Many grind the root to pair with chocolate for desserts, or mix it into simmering braises to add layers of depth to second courses.

White Chocolate and Cayenne Pepper Truffles

TARTUFI AL CIOCCOLATO BIANCO E PEPERONCINO

8 ounces (226 g) white
chocolate

¼ cup (59 ml) heavy cream

⅛ teaspoon cayenne pepper

¼ cup powdered sugar

The Calabrian panache with spice is even seen in desserts. At Dattilo these white chocolate truffles are enhanced with a pinch of dried hot pepper that adds a lingering fiery kick.

1. Roughly chop 4 ounces of the white chocolate and set aside in a large heatproof bowl.

2. Bring the cream to a simmer in a small saucepan over medium-high heat. Pour the cream over the chocolate and let sit for 2 minutes. Stir the cream and chocolate together until smooth and shiny. Stir in the cayenne pepper. Put the bowl in the freezer until the chocolate is firm and completely chilled, about 2 hours.

3. Chop the remaining chocolate and place it in a bowl set over a pot of barely simmering water until melted. Set aside to cool.

4. Scoop out 1 teaspoon of the chocolate-pepper mixture and roll into a ball. Place the truffle on a small baking sheet lined with parchment paper and finish making the remaining truffles.

5. Sift the powdered sugar into a small bowl. Dip each truffle into the melted white chocolate and set back on the parchment paper to firm up a bit, then roll the truffles in the powdered sugar. Arrange on a clean plate and refrigerate until the outer coat of chocolate is fully set, about 30 minutes.

MAKES 14 TRUFFLES

LE PUZELLE

Off a winding road that leads down from the medieval village of San Severina, past free-range animals grazing openly in the mountainous rocky outcroppings, Le Puzelle agriturismo awaits those seeking solace in the Calabrian countryside. Surrounded by cypress trees, olive groves, and citrus, the farm is an oasis of tranquility set in a panorama of landscape unchanged since the farm's founding in the early 1800s. Low-lying stone buildings with burnt orange moss-covered terra-cotta roofs line the property, and a small pool is hidden among olive trees.

After retiring from a banking career, Vincenzo Bisceglia, with the help of his companion Elvira, sought to continue his family's farm on a smaller scale, scaling back production to open an agriturismo. The couple restored the original hayloft into bedrooms and a large open dining room that opens to a spacious stone patio whose view extends to the Ionian Sea a few miles away. Guest rooms are simple but the restaurant has a rustic bourgeois feel to it, with an ornate open fireplace at its entrance, beneath chestnut beams and a vaulted ceiling, and large open windows that look out onto the courtyard. Ani-

Le Puzelle farmhouse

mals are no longer reared at the farm; instead olive oil has become Le Puzelle's main agricultural focus, in addition to a small crop of citrus fruits. These two ingredients highlight meals at the restaurant, whose long-term chef Salvatore Vona has established a reputation for his commitment to cooking with only seasonal local ingredients.

A few miles from Salvatore's kitchen sits the Sila National Park of Calabria. Home to snowcapped mountains and pine chalet villages, the Sila produces some of the region's most prized foods. Black pigs, an abundance of mushrooms (most importantly the prized porcini), and wild herbs are all specialties of the park, and Salvatore uses them copiously in his recipes. Over the ten years Salvatore has cooked at Le Puzelle, he has created a plethora of original dishes, which he has detailed in three cookbooks sold at the farm. A few of Le Puzelle's favorite dishes include a whole fresh ham from the hind leg of the black pig, roasted with herbs and white wine; grilled veal chops with a fresh porcini sauce; homemade maccheroni with a spicy sausage and vegetable ragù; and a neon green digestive liquor made from bay leaves. Locals often fill

Wild mushrooms from the Sila National Park

the tables, and Le Puzelle has become known as a destination for private events. Salvatore's years of culinary experiences has led him to teaching part-time at a local professional culinary school. Many of his students have come to apprentice at the agriturismo, taking with them the shared philosophy of Salvatore, Vincenzo, and Elvira, of showcasing Calabria's best native ingredients to produce traditional and inventive dishes true to the region.

Marinated Mushrooms

FUNGHI SOTT'OLIO

The crisp, damp fall air of the Sila park creates the perfect environment for wild mushrooms to flourish. Makeshift markets are set up throughout the park, with foragers displaying their findings in wooden boxes laid out on the street or in wicker baskets out of the back of their cars. Le Puzelle preserves this bounty with an olive oil marinade that guests enjoy until the springtime arrives, bringing a new growth of fresh mushrooms to the table.

1. Stem the mushrooms and trim them into ½-inch pieces.

2. Put the water, vinegar, bay leaves, and 1 tablespoon of salt into a 6-quart pan over medium-high heat. Bring to a boil, add the mushrooms, and cook at a vigorous simmer until the mushrooms are just tender but still retain some bite, 10 to 12 minutes. Drain the mushrooms and then spread them out on a sheet tray lined with a clean kitchen towel. Discard the bay leaves. Let cool completely, 2 to 3 hours.

3. Bring a large pot of water to a boil over high heat. Add either two pint ball jars or one quart ball jar along with their lids and boil for 10 minutes to sterilize. Remove the jars and lids from the water and let dry. Pack the mushrooms into the jars along with the chilies, fennel, carrot, celery, and mint; add oil just to cover. Let sit for 2 hours, and then close with the lid.

1 pound (453 g) mixed mushrooms, such as shiitake, oyster, cremini, maitake, and trumpet

6 cups water

½ cup (118 ml) white wine vinegar

2 bay leaves, preferably fresh

Kosher salt

2 dried small red chile peppers

½ teaspoon fennel seeds, lightly crushed

½ cup thinly sliced carrot

½ cup thinly sliced celery

8 small tender mint leaves

Extra-virgin olive oil, as needed to fill the jars

Ziti with Calabrian Sausage and Vegetable Sauce

MACCHERONI CON SUGO CALABRESE

This simple vegetable-and-sausage ragù is quick to prepare and makes for a nutritious and filling pasta dish any night of the week.

2 tablespoons (30 ml) extra-virgin olive oil

6 ounces (170 g) Italian sausage, cut into ¼-inch pieces

1 red onion, cut into small dice

1 red bell pepper, seeded and cut into small dice

1 small zucchini, quartered and then cut into ¼-inch-thick slices

1 small eggplant, cut into small dice

Pinch of crushed red pepper flakes

Kosher salt

1 28-ounce (793 g) can chopped tomatoes

1 pound (453 g) ziti

1. Heat 1 tablespoon of the oil in a 12-inch straight-sided sauté pan over medium-high heat. Add the sausage and cook until browned all over, 5 to 7 minutes. Transfer to a plate with a slotted spoon. Pour off all but a thin layer of the fat. Add the remaining 1 tablespoon of oil to the pan along with the onion, bell pepper, zucchini, eggplant, pepper flakes, and ½ teaspoon of salt. Cook, stirring occasionally, until the vegetables are tender and lightly browned, 6 to 8 minutes. Return the sausage to the pan and add the tomatoes. Bring the sauce up to a simmer, reduce the heat, and cook until the flavors have melded and the sauce has thickened, 25 to 30 minutes. Season to taste with salt.

2. Bring a large pot of well-salted water to a boil. Drop in the pasta and cook until al dente, according to package instructions. Drain the pasta well and toss immediately with the sauce. Serve immediately.

SERVES 5

Spicy Calabrian Chicken
POLLO CASARECCIO

Packed with earthy rosemary flavor, a nice hit of spice, and just the right amount of sauce this chicken dish makes for a great weeknight meal.

SERVES 4

1. Season the chicken generously with salt and pepper. Heat the oil in an 8-quart heavy-duty pan or Dutch oven over medium heat, until shimmering. Add the chicken pieces skin-side down and sear until golden brown, about 5 minutes. Flip over and brown on the other side, 3 to 4 minutes. Transfer to a plate.

2. Raise the heat to high and then pour in the wine and stir with a wooden spoon, scraping any browned bits stuck to bottom of pan. Reduce the wine by half. Add the pepper flakes, tomatoes, and rosemary, and bring the mixture to a simmer. Return the chicken to the pan, cover, reduce the heat, and simmer gently until the chicken is fork-tender, 30 to 35 minutes.

1 3-pound (1.36 kg) chicken, cut into 4 pieces

Kosher salt and freshly ground black pepper

2 tablespoons (30 ml) extra-virgin olive oil

½ cup (118 ml) white wine

¼ teaspoon crushed red pepper flakes

1 28-ounce (793 g) can whole plum tomatoes, crushed by hand

2 sprigs rosemary

Spicy Calabrian Chicken

251

Asparagus and Porcini Mushroom Lasagna
LASAGNE AGLI ASPARAGI E FUNGHI PORCINI

Thin spears of wild asparagus sprout from feathery fronds that proliferate the countryside. Paired with porcini mushrooms, this lasagna dish is a refreshing taste of spring.

12 ounces (340 g) boxed lasagna noodles

2 tablespoons (30 ml) extra-virgin olive oil

1 clove garlic, smashed

1 bunch pencil asparagus, trimmed and cut into ¼-inch pieces

7 ounces (200 g) porcini mushrooms, cut into ¼-inch pieces

Kosher salt

½ cup (118 ml) white wine

1 recipe béchamel (see "Basic Recipes")

1½ ounces (45 g) ham

1½ cups grated provolone

½ cup grated Parmigiano Reggiano

1. Bring a large pot of salted water to a boil. Prepare a large bowl of ice water. Slip the noodles, two or three at a time, into the boiling water and cook them until they're al dente, following the instructions on the box. Scoop the noodles out of the water with a large wire skimmer and slide them into the ice water to stop the cooking. When they're cool, layer them between clean dish towels until you're ready to assemble the lasagna.

2. Heat the oil in a 12-inch skillet over medium heat until shimmering. Add the garlic and cook until the oil is fragrant, about 1 minute. Discard the garlic. Add the asparagus, mushrooms, and a pinch of salt; cook until the vegetables are tender and beginning to brown, 4 to 6 minutes. Pour in the white wine and reduce to dry, 3 to 4 minutes. Add ¼ cup of water and reduce by half. Season to taste with salt.

3. Position a rack in the center of the oven and heat the oven to 375°F (190°C).

4. In a large bowl, mix together the vegetables, béchamel, and ham. In a baking dish that's about 9 × 12 × 3, ladle ¼ cup of the sauce on the bottom. Then cover with a slightly overlapping layer of cooked noodles, cutting them as needed to fill the gaps. Spread ½ cup of sauce over the first layer of noodles. Sprinkle with ½ cup of provolone and 1 tablespoon Parmigiano. Add a new layer of noodles, overlapping them slightly. Repeat the layers as instructed above, until all of the filling ingredients are used, to make a total of four layers. For the final layer, put a thin smear of sauce over the top and sprinkle with ¼ cup of Parmigiano.

5. Bake in the oven until the sauce is bubbly and the top is browned and crispy, 35 to 40 minutes. Let cool slightly before serving.

SERVES 6

Picking greens

Chocolate-Dipped Chocolate Cookies

SUSUMELLE

4 cups (500 g) all-purpose flour

3/4 cup granulated sugar

2 tablespoons Dutch processed
cocoa powder

3/4 teaspoon baking powder

1/2 teaspoon baking soda

1/4 teaspoon ground cinnamon

Pinch of ground cloves

12 ounces (340 g) bittersweet
chocolate, finely chopped

3/4 cup mild honey, such as clover

3/4 cup (175 ml) water

1 teaspoon orange zest

1 teaspoon lemon zest

*These triple-chocolate, spice-scented cookies are a Calabrian Christmastime spe-
cialty. The soft crumbly texture and sweet, spicy, floral, chocolate-filled goodness
make them too good to reserve for only one time of year.*

1. Position a rack in the center of the oven and heat the oven to 350°F (180°C).
   Line two baking sheets with parchment paper.

2. In a stand mixer fitted with the paddle attachment, mix together the flour,
   sugar, cocoa powder, baking powder, baking soda, cinnamon, and cloves on
   low speed. Add 4 ounces of the chocolate, the honey, water, and orange and
   lemon zests; mix on medium speed until a dough forms.

3. Scoop out a piece of dough the size of a golf ball and roll it into a ball. Place
   the ball on one of the prepared baking sheets and flatten with your hand.
   Finish with the remaining dough.

4. Bake one sheet at a time until puffed and somewhat firm to the touch,
   12 to 15 minutes. Let cool completely on a wire rack.

5. Meanwhile, put the remaining 8 ounces of chocolate in a bowl set over a
   pot of barely simmering water until melted. Let the chocolate cool slightly.
   Dip the rounded side of each cookie into the melted chocolate and set on a
   rack until the chocolate is set, about 1 hour.

MAKES ABOUT 40 COOKIES

CHAPTER 11
PUGLIA (APULIA)

TORREVECCHIA

Puglia's wide-open flat terrain and temperate climate of sunny days and little rain provide an ideal environment for farming. Historically the region was a destination for shepherds from surrounding mountainous regions, who in the winter months led their animals down into the plains to graze on the fresh grass and herbs. Over time this migration gave way to the development of huge plots of land planted with fruits, vegetables, olive groves, and vineyards. These farms now span across the entire region, and their products are shipped throughout Italy and beyond into Europe—Puglia is recognized as one of Italy's most profitable agricultural regions.

Yet the sun-drenched landscape is currently undergoing another transformation, focusing on something politicians are hoping will launch Puglia into the forefront of green energy: solar power. Puglia is shedding some of its farming roots and ripping out the endless fields of vegetables and twisted vines, once the backbone of the economy, and replacing them with the sleek metal of solar panels that glisten in the blazing sun. In the heart of this radical reform, the Torrevecchia farm and agriturismo defiantly maintains an agricultural lifestyle. The four Manni brothers own and run Torrevecchia farm as it

Greenhouse roses

PUGLIA (APULIA)

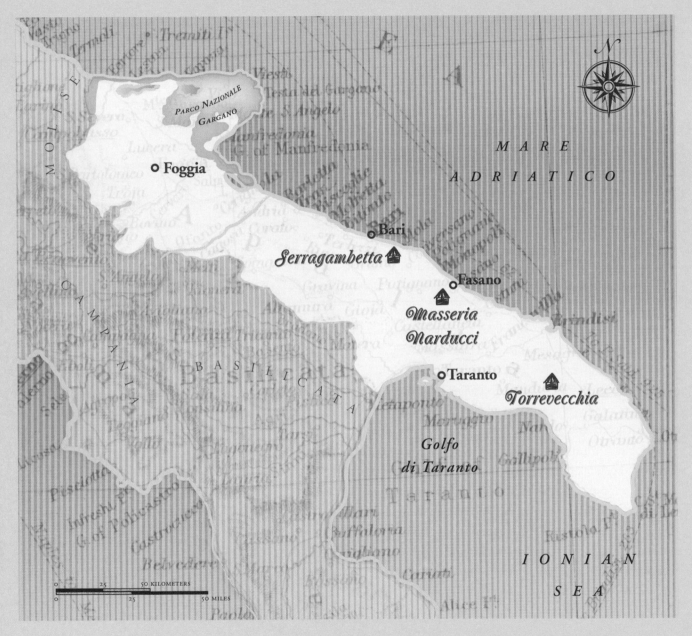

has been done since the fourteenth century. In a sprawling setting reminiscent of the large American plantations of the 1700s, the Manni's produce an astounding number of products, ranging from vegetables, wine, olive oil, meat (beef, veal, pork, and lamb), and cheese, to a separate greenhouse that grow roses. Tucked away in a secluded setting, the agriturismo distances itself from the farm's activities, offering placidity and leisure in the beautifully restored barns of the original farmhouse.

In the Torrevecchia kitchen the mother-daughter chefs use the bounty of their land as the basis of their seasonal menu, which has won numerous awards from Italian food and travel publications. Hot and cold assorted appetizers are rich in variety, with fresh and aged cheeses produced on the farm making up the majority of the offerings. At the crack of dawn, freshly pumped cow's milk arrives in the small cheesemaking laboratory adjacent to the restaurant. The small selections of handcrafted cheeses include a fresh and semi-aged table cheese, ricotta, and a fermented ricotta known as ricotta forte. A plate of deep-fried dough stuffed with vegetables, tiny meatballs, and potato croquettes—crispy on the outside and velvety on the inside with a subtle nuance of fresh mint—starts things off. An assortment of tomato- and onion-stuffed focaccia follows, joined by sautéed wild chicory with pecorino. Pasta fashioned from Tor-

revecchia's own semolina flour is handmade and sauced with tomatoes, capers, anchovies, olives, chickpeas, and seafood taken from the nearby Ionian Sea. Meats are simpler preparations; a favorite veal dish is done pizzaiolo (pizza maker) style, topped with a tomato sauce. Although not classically Puglian, veal saltimbocca, made with Torrevecchia's prosciutto, offers a salty taste of the farm's pork. Summers and weekends find the agriturismo packed with vacationing Italians enjoying the hard work and efforts of the families of Torrevecchia, who are determined to continue making a living off their land in the significantly changing Puglian countryside.

Trulli

Potato Croquettes with Mint
CROCHETTE DI PATATE CON MENTA

Mint adds a very nice note to these potato croquettes. Feel free to play around with the recipe, adding other herbs that suite your palate.

4 Yukon Gold potatoes (1¾
    pounds/820 g total), peeled and
    quartered
Kosher salt
2 large eggs
⅓ cup grated pecorino
¼ cup grated Parmigiano Reggiano
3 tablespoons finely chopped mint
Freshly ground black pepper
½ cup bread crumbs
Canola oil for frying

1. Put the potatoes in a 4-quart pot and add enough water to cover by 1 inch. Add 2 teaspoons of salt and bring to a boil over medium-high heat. Reduce the heat to maintain a simmer and cook until the potatoes are tender when pierced with a skewer or fork, 10 to 12 minutes. Drain the potatoes, put them back in the pot, and cook over medium-low heat to dry them out, about 30 seconds. Pass the potatoes through a ricer into a large bowl. Add the eggs, grated cheeses, mint, and a few grindings of pepper; stir well to combine.

2. To form the croquettes, scoop about 1 tablespoon of the potato filling and shape it into a 1-inch-long oval. Put the bread crumbs in a shallow bowl. Roll the croquettes in the bread crumbs to coat evenly and then transfer to a baking sheet lined with parchment. Continue with the remaining croquettes. Place the baking sheet in the refrigerator until the croquettes are firm, about 1 hour.

3. Heat ½ inch of canola oil in a 12-inch cast-iron skillet over medium heat until it reaches 350°F (180°C) on a candy thermometer. Add six to eight croquettes to the oil and cook, turning once, until a they're deep golden brown all over—4 to 6 minutes total. Transfer the croquettes with a slotted spoon to a plate lined with paper towels. Continue frying the remaining croquettes in the same manner, adjusting the heat as necessary to maintain the temperature of the oil at 350°F. Serve warm.

MAKES 39 CROQUETTES

Potato Croquettes with Mint

Dandelion Greens with Guanciale, Pecorino, and Black Olives

CICORIA CON GUANCIALE, PECORINO, E OLIVE

In autumn wild chicory grows in the fields surrounding Torrevecchia. This bitter green appears on the menu in myriad ways, but one of our favorites was this flavorful side dish. If chicory or dandelion greens are unavailable, substitute escarole.

1. Heat ½ tablespoon of the oil in a 12-inch skillet over medium heat. Add the guanciale and cook until it has rendered its fat and is just beginning to crisp, 2 to 3 minutes. Remove from the pan with a slotted spoon, and drain on a paper-towel-lined plate. Pour off the fat from the pan.

2. Heat the remaining 1 tablespoon of oil in the pan over medium heat until shimmering. Add the dandelions and a pinch of salt and cook, covered, until the greens have wilted, about 5 minutes. Add the guanciale back to the pan along with the pecorino cheese, cover, reduce the heat to medium-low, and cook until the greens are very tender, 8 to 10 minutes. Add the olives to the pan, season to taste with salt and pepper, and stir to combine. Serve on a platter.

SERVES 4

1½ tablespoons (20 ml) extra-virgin olive oil

2 ounces (60 g) guanciale or pancetta, cut into fine dice

1 large bunch dandelion greens, trimmed and roughly chopped

Kosher salt

¼ cup grated pecorino cheese

¼ cup black olives, such as niçoise, pitted and quartered lengthwise

Freshly ground black pepper

Wild Chicory

Cavatelli with Navy Beans and Cherry Tomatoes

CAVATELLI CON FAGIOLI E POMODORINI

Throughout southern Italy beans are cooked in terra-cotta urns placed directly into a burning fire. The porous pot regulates the cooking temperature, rendering legumes tender and creamy.

4 ounces (115 g) dried navy beans, soaked in water overnight

3 small cloves garlic, smashed

2 bay leaves

1 stalk celery, cut into 3-inch pieces

Kosher salt

2 tablespoons (30 ml) extra-virgin olive oil

12 ounces (340 g) cherry tomatoes, halved

½ cup crushed tomatoes

1 pound (453 g) boxed cavatelli

1 tablespoon chopped flat-leaf parsley

1. Drain and rinse the beans and drain again. Add the beans, 2 cloves of the garlic, the bay leaves, and the celery to a 4-quart pot and cover with water by ½ inch. Bring to a boil over medium-high heat. Reduce the heat to maintain a gentle simmer and cook until the beans are creamy and tender, 30 to 45 minutes depending on the quality of the bean. Season with a generous pinch of salt and cook for 2 more minutes. Drain the beans and discard the garlic, bay leaves, and celery.

2. Heat the oil in a 4-quart pan over medium heat with the remaining 1 clove garlic and cook until the garlic is golden brown. Discard the garlic, raise the heat to medium-high, and add the cherry tomatoes and a pinch of salt. Simmer the tomatoes until they just begin to break down, 5 to 8 minutes. Stir in the crushed tomatoes and continue cooking until the sauce thickens, 8 to 10 minutes. Season to taste with salt and then set aside.

3. Bring a large pot of well-salted water to a boil over medium-high heat. Drop in the pasta and cook until al dente, following the package instructions. Reserve ¼ cup of the pasta water and then drain the pasta well.

4. In a large bowl, toss the pasta with the tomato sauce. Mix in the beans and add some of the reserved pasta water, 1 tablespoon at a time, if the sauce seems dry. Add the chopped parsley and season to taste with salt and pepper. Serve in individual shallow bowls

SERVES 6

Beans cooking in terra-cotta urns

Fried Dough with Honey

PITTULE

2 cups (234 g) all-purpose flour

1 teaspoon fine sea salt

3/4 teaspoon active dry yeast

1 1/3 (315 ml) cups warm water

Canola oil, for frying

Honey, as needed for serving

These addictive little fritters can be made with a host of different fillings, like sautéed finely chopped red onion, diced cooked tomatoes with capers, or grated pecorino cheese. Or for a sweet treat, you can leave out the filling, drizzle honey over the warm fritters, and serve with a glass of dessert wine.

1. In a large bowl, mix together the flour, salt, and yeast. Gradually mix in the water and stir until you have a very wet dough, adding more water if necessary. The dough should be loose, but not pourable like a batter. Cover the bowl with a kitchen towel and let it sit at room temperature until it has doubled in volume, about 2 hours.

2. In a 4-quart pot, heat 1½ inches of canola oil over medium-high heat until it reaches 360°F (182°C) on a candy thermometer. Have ready a small bowl of cold water and two dinner spoons. Dip the dinner spoons in the water. Remove the spoons from the water and with one of them, scoop up a mound of the batter, then use the other to slide the dough into the hot oil. Add about five more spoonfuls of batter into the oil and fry until the dough puffs up and becomes a light golden brown, 2 to 3 minutes. Using a slotted spoon, transfer the fritters to a plate lined with paper towels. Continue frying the remaining fritters in the same manner.

3. Serve warm drizzled with honey.

YIELD - ABOUT 3 DOZEN

Fried Dough with Honey

FRIED OLIVES

Aside from oil, Puglia's olive trees supply an endless bounty of fruit, whose different colors, shapes, and sizes share the table with numerous antipasti. When taken straight from the tree, most olives are inedible, with an incredibly bitter and astringent taste; that is subdued by a long soak in a saltwater brine. In Puglia, however, there exists a particular type of tree that produces large, meaty black olives, which are picked straight from the tree and fried in extra-virgin olive oil (naturally). After a quick dip into the shimmering oil, they are tossed with bay leaves, diced hot peppers, and a generous sprinkling of salt, and are served warm.

SERRAGAMBETTA

In a region of recent discovery on the tourist map, where five-star resorts and golf courses are beginning to plant their seeds of inevitable change, the agriturismo of Serragambetta, owned by mother Nina and son Domenico Lanera, stands deeply rooted in old-fashioned, genuine hospitality. The epicenter of Serragambetta lies within the white limestone bricks of its kitchen. The enchantment of the room is enhanced with copper pans that hang from the walls, terra-cotta urns adorning the fireplace mantel, mismatched jars of dried herbs that fill the worn wooden shelving, and antique pine cabinets with cracked marble countertops. These all help to create an ambience of warmth and comfort, but it is Nina, affectionately known to all as Zia (aunt) Nina, who brings the kitchen to life. All day long the kitchen bustles with organized commotion. In addition to preparing the night's family-style dinners, the kitchen maintains a rigorous schedule of making jams from the plum, cherry, orange, quince, and fig trees around the property. The doors are always open for guests to stop by, sit for a while in front of the fire, drink some tea, and to see what is simmering away on the stove. Domenico comes and goes from his work in the fields tending the olive and fruit trees, igniting even more life in the kitchen with his joking and playful banter—always directed at his mother. This congenial family atmosphere has brought great success to Serragambetta since its opening over twenty-five years ago, appealing to vacationers seeking a genuine experience.

Preserving jam

The agriturismo's long tenure coupled with its recognition as one of the area's original and premier farm vacations has brought honorable mentions in trusted international guidebooks. The farm is so popular that rooms can be hard to come by, as guests travel here year-round from all corners of the world. Nina's four-course Pugliese meals emphasize Domenico's organically grown vegetables. Beans cooked in an earthenware pot nestled among the embers in the kitchen's fireplace is a staple in her repertoire, often paired with handmade pasta. Area fishermen stop by with their catches, and mussels are a favorite, often cooked with zucchini for a delicious first course. Roasted chicken and rabbit cooked on a bed of rosemary branches are mainstay second courses, perfumed with the piney scent. While Nina commands the kitchen, dishing out generous portions of antipasti, pastas, meats, and desserts, Domenico helps out by firing up the brick oven. A purist, he makes tomato focaccia recipe by mixing boiled potatoes into the dough—a mandatory ingredient when making the classic Puglian flatbread. The starchy vegetable creates a deliciously crispy and chewy crust.

Guests are seated together at one long wooden table that stretches across the dining room, resembling a very informal dinner party. What begins as a quiet room full of strangers gradually comes to life as inhibitions are warmed by good food and wine, and new friendships made. A real treat awaits guests luckily enough to eat at Serragambetta when Nina's ninety-seven-year-old mother comes to spend the day making fresh pasta. Her deft and nimble

Zia Nina in the kitchen

hands effortlessly shape orecchiette and cavatelli in rapid succession, while she shares stories of Puglia's past during times of war and rationing. Her wit and infectious charm stir awe at the depths of Italy's culinary history. Her mere presence is symbolic of why Serragambetta is such a special place to savor the pleasures of Puglia's countryside.

Focaccia

Adding boiled and riced potatoes to the dough is the secret to making authentic soft and billowy Pugliese focaccia. Toppings can include just about anything, but cherry tomatoes with a hint of rosemary and salt is the ultimate classic.

For the dough:

6 cups (700 g) all-purpose flour

1 medium Yukon Gold potato, boiled and passed through a ricer

1 tablespoon kosher salt

2 teaspoons active dry yeast

¼ cup (59 ml) extra-virgin olive oil, plus extra as needed

2¼ cups (532 ml) tepid water

For the topping:

1 pint cherry tomatoes, quartered

2 teaspoons chopped rosemary

To assemble:

2 tablespoons (30 ml) extra-virgin olive oil

1 tablespoon kosher salt

1. **Make the dough:** In a stand mixer fitted with the dough hook, mix together the flour, potato, salt, and yeast. Gradually add the oil, then mix in the water. Knead the dough on low speed until it comes together and is smooth and elastic, 8 to 10 minutes. Cover and let rise until doubled in volume, 1 to 1½ hours.

2. Moisten your fingers with a bit of oil and then spread the dough out onto a sheet tray to fill the entire tray in an even layer. Cover with a damp towel and let rise until doubled in volume, about 45 minutes.

3. **Make the topping:** In a medium bowl, toss the tomatoes with rosemary.

4. **To assemble,** position a rack in the center of the oven and heat the oven to 425°F (220°C).

5. Uncover the dough, drizzle with the oil, and press down and dimple the dough with your fingers. Spread the tomatoes and rosemary evenly over the dough and sprinkle with salt.

6. Bake in the oven until the focaccia is a deep golden brown, 45 to 50 minutes. Let cool slightly on a wire rack. Transfer the focaccia to a cutting board, cut into pieces, and serve.

Pugliese focaccia

Fresh Pea and Egg Drop Soup

STRACCIATELLA CON PISELLI

This Italian-style egg drop soup transforms a few ingredients into something delicious. It can also be made with zucchini, asparagus, or spinach.

1. Heat the oil in a 4-quart pot over medium heat. Add the onion and a pinch of salt and cook until tender and translucent, 5 to 7 minutes. Add 4 cups of water and 2 teaspoons of salt to the pot and bring to a boil. Add the peas and simmer gently, partially covered, until the peas are very tender, about 30 minutes. With an immersion blender (or in a blender), puree half of the soup.

2. Beat the eggs in a medium bowl, and then mix in the cheese, parsley, pinch of salt, and a few grinds of black pepper. Add the egg mixture to the pot and immediately remove from the heat. Stir the soup with a wooden spoon to break up the egg into pieces. Season the soup to taste with salt and pepper and serve in individual soup bowls.

SERVES 6

2 tablespoons (30 ml) extra-virgin olive oil

1 medium yellow onion, cut into fine dice

Kosher salt

3 cups shelled English peas

2 large eggs

3 heaping tablespoons grated pecorino cheese

1 tablespoon finely chopped flat-leaf parsley

Freshly ground black pepper

Domenico baking bread

Cavatappi with Black Olive Sauce

CAVATAPPI CON SALSA DI OLIVA

1 medium yellow onion, roughly chopped

3 ounces (85 g) slab bacon, roughly chopped

2 tablespoons (30 ml) extra-virgin olive oil

1 clove garlic, chopped

Kosher salt

1 14-ounce (396 g) can cherry tomatoes, crushed by hand

3 ounces (85 g) black olives, such as kalamata, pitted and halved

1 tablespoon chopped flat-leaf parsley

Freshly ground black pepper

12 ounces (340 g) cavatappi pasta

Bacon and black olives pack a flavorful punch in this simple and tasty pasta dish. Pureeing the onion and bacon creates a meltingly creamy consistency for the sauce.

1. Put the onion in a food processor fitted with the blade attachment and puree until creamy. Transfer to a small bowl. Add the bacon to the processor and pulse until very finely chopped; transfer to a separate bowl.

2. Heat the oil and garlic in a 4-quart saucepan over medium-high heat. Cook until the garlic is golden brown, about 2 minutes. Add the onion and a pinch of salt and cook until golden brown, 4 to 6 minutes. Stir in the bacon and cook until it loses its rawness, 2 to 3 minutes. Stir in the tomatoes; reduce the heat to maintain a gentle simmer and cook, stirring occasionally, until the tomatoes thicken, about 20 minutes. Stir in the olives and parsley; season to taste with salt and pepper.

3. Bring a large pot of well-salted water to a boil over medium-high heat. Drop the pasta into the boiling water and cook according to package directions until al dente. Drain the pasta well. Toss with the sauce and serve in individual shallow bowls.

SERVES 4

Ninety-seven-year-old grandmother making pasta

Farfalle with Zucchini and Mussels
FARFALLE CON ZUCCHINE E COZZE

Surrounded by water, the heel of Italy provides an abundance of seafood, which has long played a significant role in Puglia's culinary traditions. Mussels are a particular favorite, prized for their briny, salty flavor. True connoisseurs prefer eating them raw—a testament to their freshness and the cleanliness of the region's waters. Zucchini and mussels are a perfect marriage, and Zia Nina combines the two for a delectable and quick first course.

¼ cup (59 ml) extra-virgin olive oil

2 small zucchini, cut into match-sticks

Kosher salt

1 clove garlic

1½ pounds (680 g) small mussels, scrubbed and debearded

12 ounces (340 g) farfalle pasta

2 tablespoons chopped flat-leaf parsley

Freshly ground black pepper

1. Heat 2 tablespoons of the oil in 12-inch skillet over medium-high heat. Add the zucchini and a generous pinch of salt and cook, stirring often, until the zucchini is tender, about 5 minutes. Remove from the heat and set aside.

2. In an 11-inch straight-sided sauté pan, heat the remaining 2 tablespoons of oil with the garlic over medium heat and cook until the garlic is golden brown. Discard the garlic. Raise the heat to medium-high and add the mussels to the pan along with ½ cup of water. Cover the pan and cook until the mussels open up, 3 to 5 minutes. Remove the mussels from their shells and discard the shells. Strain the juices from the mussels through a fine-mesh sieve into a small bowl. Add the mussels to the liquid and set aside.

3. Bring a large pot of well-salted water to a boil. Drop in the pasta and cook according to the package directions until al dente. Drain the pasta.

4. Reheat the zucchini over medium-high heat. Add the pasta, the mussels, and their juices; toss to combine. Add the parsley and season to taste with salt and pepper. Serve in individual shallow bowls.

SERVES 4

Chocolate Bundt Cake with Orange Marmalade Ganache Glaze

CIAMBELLA DI CIOCCOLATO CON MARMELLATA D'ARANCIA

Bundt cakes are very popular desserts in Italy, since they are quick to put together and are a tasty treat any time of day. The orange marmalade enhances the bitterness of the chocolate and helps offset the sweetness of the cake.

1. **Make the cake:** Position a rack in the center of the oven and heat the oven to 350°F (180°C). Butter a 9-inch bundt pan.

2. In a large bowl, beat together the eggs and sugar with a handheld electric mixer on medium speed until pale yellow and thickened, 2 to 3 minutes. In a small bowl, stir together the flour, baking powder, and baking soda. Combine the milk and oil in a liquid measuring cup.

3. Beat half of the flour mixture into the bowl with the eggs until just combined. Add half of the milk mixture and mix until combined. Add the remaining flour, mix, and then add the remaining milk. Stir in the melted chocolate and pour into the prepared bundt pan.

4. Bake in the oven until the cake has risen and a toothpick inserted in the center comes out clean, 40 to 45 minutes. Transfer to a rack and let cool for 5 minutes, then invert the cake onto a plate. Let cool completely.

5. **Make the glaze:** Put the chocolate and butter in a bowl set over a pot of barely simmering water, and cook, stirring occasionally, until melted. Remove from the heat and let cool slightly. In a mini food processor, combine the orange marmalade and the sugar and process until smooth. Stir the orange marmalade into the chocolate. Drizzle the glaze over the cake and let set for 30 minutes before serving.

SERVES 10–12

For the cake:

Unsalted butter, as needed for the pan

5 large eggs

1½ cups granulated sugar

2⅓ cups (310 g) all-purpose flour

1¼ teaspoons baking powder

1 teaspoon baking soda

½ cup (118 ml) whole milk

½ cup (118 ml) vegetable oil

7 ounces (200 g) bittersweet chocolate, melted and cooled

For the glaze:

3 ounces (85 g) bittersweet chocolate

1 tablespoon unsalted butter

¼ cup orange marmalade

1 tablespoon granulated sugar

Almond Croccante

CROCCANTE DI MANDORLE

8 ounces (226 g) almonds, chopped

1¼ cups granulated sugar

This classic Pugliese sweet snack is just as quick to make as it is to disappear.

1. Put the almonds in a heavy-duty skillet with the granulated sugar. Set the skillet over medium-low heat and cook, stirring continuously, until the sugar begins to caramelize and the almonds are nicely toasted, 5 to 8 minutes.

2. Pour the mixture out onto a rimmed baking sheet lined with parchment paper or onto a lightly dampened marble surface. With the back of a chef's knife, spread the mixture out into an even layer. Let cool slightly, then cut the brittle into squares.

SERVES 8 TO 10

RICOTTA FORTE

Across Puglia a slightly fermented, past-its-prime ricotta is considered a regional delicacy. Ricotta forte has an assertive, piquant, and pungent flavor, an acquired taste adored by locals. While regular ricotta has a buttery, creamy consistency, ricotta forte gets its tangy bite from its three to four months spent in terra-cotta containers left in a cool, damp place, which get salted and stirred by hand every two days. Over time the cheese darkens from milky white to a caramel hue and develops a strong sour smell. The spreadable soft cheese can be eaten atop grilled slices of bread with olive oil as a bruschetta, or added to a tomato sauce and tossed with a unique type of pasta called maritate, which combines orecchiette (little ear shapes) and casarecce (2-inch-long twists). At Serragambetta, a large glass jar of the ricotta forte placed on the table with homemade bread awaits guests willing to sample this classic Puglian ingredient.

Almond Croccante

NARDUCCI

The knotted and contorted trunks of Puglia's ancient olive trees are living sculptures of history. Their fruit-bearing branches have thrived for hundreds of years, providing the region with ripe green and black olives. The groves that extend to the sea around the Narducci agriturismo have been designated as a regional park; forbidding their removal and preserving the area's legacy for their production of extremely high-quality extra-virgin olive oil. Winding narrow roads meander through the countryside, flanked by white limestone walls that contain endless green vegetable fields of broccoli rabe, swiss chard, and lettuces in burnt-red-hued earth, between rows of massive olive trees, all set against a backdrop of a bright blue Puglian sky. These contrasts in the colors are visually breathtaking, and the huge stone farmhouses tucked into the landscape only add to the visual charm of the area.

Here the Narducci family has been pressing oil from some of the oldest and most prized trees for generations. They converted their working farmhouse into an agriturismo in the late 1990s, scrupulously restoring it to maintain all of the characteristics of the Pugliese masseria (farmhouse). The impeccably white stucco building is surrounded by lush gardens. Palms and cacti, red peppercorns, figs, citrus trees, and olive trees are dispersed throughout the estate. The original masseria housed the olive mill on the ground floor. The open stone-vaulted room has been transformed into the agriturismo's dining room, decorated with vintage farming equipment. The space is airy and casual, and a perfect place to linger over a classic Pugliese feast.

Jolly and always ready with a joke, Giuseppe Narducci manages his family's farm with a keen sense of humor. He declares himself the farm's official food taster, and his slight paunch adds to his jovial demeanor, which enhances an unforgettable dining experience. Dishes arrive in bountiful quantities, prepared with love and precision by his mother Maria. Kicking off each meal are an apparently infinite number of antipasti that come piled high on a tray. With barely enough room on the table to accommodate all of the tiny terra-cotta bowls and plates, diners are presented with an array of Puglian appetizers, including fried black olives, batter-fried vegetables, wheatberry salad, zucchini and eggplant frittata, stewed tripe, and gratinéed mushrooms, to name a few. If guests are still hungry after all this, primi may include house-made orecchiette with broccoli rabe; tagliatelle with artichoke hearts, ground pork, and veal; or an open peapod-shaped pasta with local mushrooms (cardoncelli) and ground sausage. A classic second course not for the squeamish are turcinieddi, which combine lamb heart, lung, and liver stuffed into intestines. These are roasted, sliced, and served alongside other meat dishes such as rabbit or veal involtini. The evening concludes with a refreshing platter of fruit from the garden and an assortment of almond cookies. In the familiar ambience of Narducci, the dining room emanates comfort; it's like eating at Grandmother's house. No wonder guests stay long after their plates have been cleared, talking late into the night.

Olive tree in red soil

POMODORO REGINA

Little rain falls in the dry and arid climate of Puglia, and access to fresh water has always been a thorn in the side for the regions' farmers. Basins were often built on rooftops to catch the small amounts of rainfall, which was used sparingly for crops withering beneath the blistering sun. In the 1950s, however, underground rivers were detected running deep beneath the earth. Wells were drilled, drastically improving growing conditions for Puglian farmers. After a few grow-

ing seasons, it was discovered that the water affected certain vegetables; tests revealed that the rivers contained high levels of salinity. One such vegetable was a local cherry tomato, called the Pomodoro Regina, which developed thick skin from absorbing the brackish water. Locals found that when the tomatoes were strung and hung indoors in a cool environment, their new thicker skins allowed them to endure entire winters without spoiling. In the days before greenhouses and hothouse tomatoes, this provided a welcome freshness in a diet otherwise heavy on canned tomatoes and hearty winter greens. At Narducci and in homes throughout Puglia, a favorite appetizer or snack involves splitting the tomatoes open and squeezing their juices over bruschetta or freselle.

Bread, Green Bean, and Mozzarella Frittata

FRITTATA AL PANGRATTATO CON FAGIOLINI E MOZZARELLA

This simple antipasto can be made with any fresh vegetables. The addition of bread crumbs makes this recipe unique and helps give substance to the egg-based appetizer, which is best eaten at room temperature.

1. Put the bread crumbs in a small bowl and dampen with 2 tablespoons of water. Let sit at room temperature for 30 minutes.

2. Bring a 4-quart pot of salted water to a boil over high heat. Add the green beans, return to a boil, and cook until just tender, 3 to 5 minutes. Drain the green beans and immediately submerge in a bowl of ice water. Drain again, dry the green beans with a kitchen towel, and then cut into ½-inch pieces.

3. In a large bowl, mix together the bread crumbs, grated cheese, parsley, mozzarella, green beans, ½ teaspoon salt, and a few grinds of pepper; stir well to combine. Add the eggs to the bowl and mix all the ingredients together with your hands until thoroughly combined.

4. Position a rack in the center of the oven and heat the oven to 375°F (190°C).

5. Lightly oil an 8 × 8 × 2-inch baking dish. Spread the bread mixture evenly into the pan and bake in the oven until a deep golden brown, about 25 minutes. Transfer to a rack and let cool slightly. Cut the frittata into small squares and serve.

SERVES 6–8

1 cup coarse bread crumbs

6 ounces (170 g) green beans, trimmed

½ cup grated pecorino cheese

2 tablespoons chopped flat-leaf parsley

4 ounces (115 g) fresh mozzarella, finely chopped

Kosher salt and freshly ground black pepper

6 eggs, lightly beaten

Gratinéed Oyster Mushrooms
FUNGHI AL FORNO

½ pound (227 g) oyster mushrooms, trimmed and separated into pieces

Kosher salt and freshly ground black pepper

¾ cup coarse bread crumbs

1½ tablespoons (22 ml) extra-virgin olive oil, plus more for the pan

2 tablespoons (30 ml) white wine

2 tablespoons chopped flat-leaf parsley

1½ teaspoons capers

Appetizers at Narducci are plentiful and varied. They typically rely heavily on vegetables that thrive in Puglia's temperate and sunny climate. Other vegetables you can use in this recipe include eggplant and tomato, zucchini and onion, and artichokes.

1. Position a rack in the center of the oven and heat the oven to 375°F (190°C).

2. Lightly oil a 9 × 13 × 2-inch baking dish. Add the mushrooms to the dish in a slightly overlapping, single layer. Season the mushrooms with ¼ teaspoon of salt and a few grindings of pepper.

3. In a medium bowl, moisten the bread crumbs with 1 tablespoon of the oil and the white wine. Add the parsley, capers, ¼ teaspoon salt, and ⅛ teaspoon pepper. Stir well to combine. Sprinkle the bread crumb mixture over the mushrooms and then drizzle the top with the remaining ½ tablespoon of oil. Roast in the oven until the bread crumbs are a deep golden brown and the mushrooms are tender, 30 to 35 minutes.

SERVES 4

Narducci dining room

Orecchiette with Broccoli Rabe

ORECCHIETTE CON CIMA DI RAPA

This dish represents the iconic Pugliese first course. At every farm we visited there, we saw fresh orecchiette made by hand. It was almost always paired with broccoli rabe, a vegetable that grows profusely in the fertile red-soiled fields of the region. The key in making this dish is boiling the pasta in the same water in which you blanch the broccoli rabe.

1. Bring a large pot of well-salted water to a boil. Add the broccoli rabe and cook until just tender, 3 to 4 minutes. Remove the broccoli rabe from the boiling water with a spider or slotted spoon and spread out on a baking sheet lined with a clean dish towel. Let this sit to steam and release moisture, about 7 minutes. Use the dish towel to gently wring the greens and get rid of any remaining moisture.

2. Add the orecchiette to the pot of boiling water and cook until al dente, according to the directions on the package. Reserve ½ cup of the pasta water and then drain the pasta.

3. Heat the olive oil in a 12-inch skillet over medium-high heat. Add the anchovies and crushed red pepper, break up the anchovies with a wooden spoon, and cook until the oil is fragrant, 1 to 2 minutes. Add the broccoli rabe and cook until the flavors have melded, 3 to 5 minutes. Add the orecchiette to the pan and toss well to combine, adding a bit of the reserved water as needed to help coat the pasta. Season to taste with salt and pepper flakes. Transfer to a large bowl and serve.

SERVES 4

Kosher salt

1 bunch (about 8 oz./226 g) broccoli rabe, trimmed and cut into 1-inch pieces

12 ounces (340 g) orecchiette pasta

3 tablespoons (44 ml) extra-virgin olive oil

2 anchovy fillets

Pinch of crushed red pepper flakes

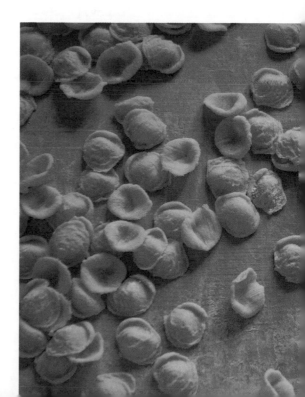

Handmade orecchiette

Freselle

Puglians love their high-quality durum wheat, making pasta, bread, and an array of baked snacks from the sandy yellow flour. During our travels throughout Puglia, a bag of crunchy ring-shaped crackers called taralli was always open in our car and fueled our drives around the countryside. We spent one afternoon at the country house of a friend who immigrated to the States a few years ago to open an Italian restaurant in New Haven, Connecticut. When we visited his family, they fired up their brick oven and taught us how to make freselle, a twice-baked super-crunchy biscuit. After being shaped into rings, they puff up during their first round in the oven, resembling American bagels. While they are still warm, they are cut in half and set back into the oven at a lower temperature, to dry out and achieve their addictive crispy bite. Try these broken up into stews, soups, or salads, or—as they do in Puglia—try eating them whole, sprinkled with chopped tomatoes and a drizzle of olive oil.

MAKES 36 HALVES

1½ pounds (680 g) semolina flour

1 tablespoon kosher salt

1 teaspoon instant yeast

2 cups (473 ml) tepid water

1. In a stand mixer fitted with a dough hook, mix together the flour, salt, and yeast on low speed until combined. Gradually add the water and mix on medium-low until the dough comes together and is smooth and elastic, 10 to 12 minutes. Cover the bowl with plastic wrap and set aside in a warm, draft-free spot until the dough has doubled in volume, about 1 to 1½ hours.

2. Turn the dough out onto a lightly floured surface. Cut off a piece of dough, about 2 ounces, and roll it out into a 9-inch-long rope that's about 1 inch thick. Connect the ends of the rope to form a circle. Transfer the circle to a lightly oiled baking sheet and finish shaping the remaining dough. Cover the freselle with lightly oiled plastic wrap and set aside to rise until doubled in volume, 45 minutes to 1 hour.

3. Position a rack in the top and bottom thirds of the oven and heat the oven to 400°F (200°C).

4. Bake the freselle until puffed and golden brown, 25 to 30 minutes. Remove from the oven and let cool slightly. Cut the freselle in half crosswise and return to the baking sheets, cut-side up. Return to the oven, reduce the temperature to 325°F (170°C), and continue baking until nicely browned and dried out, 45 to 50 minutes. Transfer to a wire rack and let cool completely.

Oven-Roasted Rabbit with Potatoes and Shallots

CONIGLIO AL FORNO CON PATATE E LAMPASCIONI

Lampiscioni are wild and bitter, and are coveted in the fall months. A traditional method of using them is by roasting them with potatoes and rabbit (or chicken). Here we use shallots, whose mild flavor pairs nicely with the delicate notes of the rabbit.

1 2-pound (900 g) rabbit, cut into 8 pieces

Kosher salt and freshly ground black pepper

12 small shallots, peeled and trimmed (halved if large)

2 small Yukon Gold potatoes, peeled and cut into ¼-inch-thick rounds

3 sprigs rosemary

1 clove garlic, thinly sliced

2 tablespoons chopped flat-leaf parsley

1 tablespoon grated pecorino

1 tablespoon (15 ml) extra-virgin olive oil

¾ cup (175 ml) dry white wine

1. Position a rack in the center of the oven and heat the oven to 425°F (220°C).

2. Season the rabbit with salt and pepper, and then arrange in a roasting pan. Evenly distribute the shallots, potatoes, rosemary, and garlic in the pan. Sprinkle with the parsley and pecorino and then drizzle with the oil. Add the wine to the pan, then roast in the oven until the rabbit is cooked through and nicely golden brown, 45 minutes to 1 hour. Transfer to a large platter and serve.

SERVES 4

Orange-Infused Baby Artichokes

CARCIOFI ALL'ARANCIA

Braising enhances the natural tenderness in baby artichokes. The bread crumbs help thicken the liquid to create a dense sauce, while citrus adds brightness to the faint earthy flavor of the artichoke.

1. Squeeze the lemon quarters into a large bowl filled with cold water and then add to the bowl. Clean the baby artichokes by removing the dark green outer leaves until only the pale, tender inner leaves remain. Trim ½ inch from the top of the artichokes, and then trim the stem end and any dark parts around the bottom. Cut the artichokes in half and add them to the bowl of lemon water.

2. Zest the orange and set aside in a small bowl. Juice the orange and add it to the bowl with the zest.

3. Remove the artichokes from the lemon water and put in a 4-quart pot. Add the orange zest and juice, parsley, oil, ½ teaspoon of salt, and a few grinds of pepper. Pour in 1 cup of cold water and then bring to a boil over medium-high heat. Reduce the heat to maintain a gentle simmer and cook until the artichokes are just tender, about 12 minutes. Sprinkle with the bread crumbs and continue cooking until the artichokes are tender when pierced with a fork and almost all of the liquid has evaporated, 12 to 15 minutes. Season to taste with salt and pepper. Transfer to a bowl and serve warm.

SERVES 4

1 lemon, quartered

8 baby artichokes

1 medium orange

1 tablespoon chopped flat-leaf parsley

1 tablespoon (15 ml) extra-virgin olive oil

Kosher salt and freshly ground black pepper

2 tablespoons bread crumbs

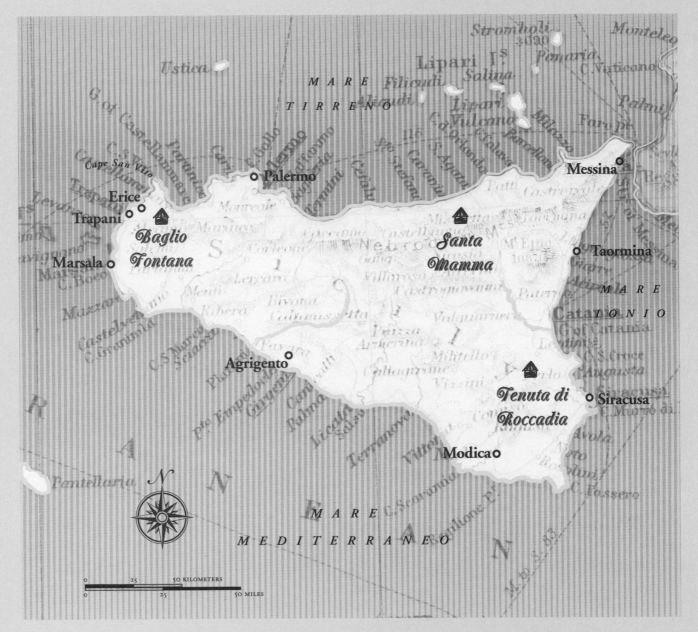

SICILIA (SICILY)

CHAPTER 12

SICILIA (SICILY)

SANTA MAMMA

A twenty-minute ferry ride across the choppy channel of the Strait of Messina marks the arrival to the gateway of Sicily—and to an entire different Italy from the shores across the shimmering water. As you drive east on autostrada E90 with the windows down, your senses become acutely heightened, awakened by floral scents, salty air, and the enchanting scenery of palm trees, bright flowers, and sparkling sunshine dancing off the water flanking the highway. The exit for Santa Mamma takes you down to sea level, offering a view of a distant crop of rocks rising from the ocean known as the Aeolian Islands. This vision alone seems to capture the essence of Sicilian island charm. You'll then begin an immediate ascent of an 11-mile mountain road pocked with deep potholes and winding curves that leads into the belly of Nebrodi National Park and into a distinct part of Sicily. Zigzagging through forests dense with cork, beech, and oak trees, open fields of olive groves, stretches of peaks and valleys, past flowing rivers and isolated stone farmhouses, with herds of free-range animals often blocking the way, the road eventually leads to a solitary spot of rare and stunning beauty: the Santa Mamma farm and agriturismo.

Alfonso Collura and his wife, Giusy, have created a rare gem for those seeking an idyllic holiday. With a

full-time career as an astrophysicist in Palermo, Alfonso wanted to reconnect with his roots and to revitalize his five-generation family farm by opening Santa Mamma as a part-time escape from the city. Today the agriturismo showcases a way of life that has remained little changed throughout Sicily's history of wars, Mafia control, banditry, and mass emigration, and Alfonso continues to work his land the way his ancestors once farmed. There are no barns at Santa Mamma, and the 1,200-plus acres are home to native black pigs, goats, sheep, horses, cows, water buffalo, and donkeys, which all live in the open fields and

Free-range cattle

wooded hamlets of the property. Walking the trails that meander around the farm, you'll encounter them roaming and grazing freely on the lush Mediterranean plants, herbs, and grasses of the landscape. To Alfonso, this represents the only proper way to raise animals.

Low-lying orange and red stone buildings, which once housed a small, sustainable farming community, have been converted into a warm and inviting agriturismo. The complex encircles an open piazza with tables to sit at and soak in the unending green mountain vista, breathing clean invigorating air fragrant with wild herbs and flowers. The village's small chapel and communal brick oven are still in use, and a restaurant and lounge reside in the main farmhouse and are decorated with family heirlooms, copper pots, and ceramic urns. Narrow wooden tables fit into the bucolic atmosphere, lending the feeling that time has stopped.

Alfonso and Giusy care deeply about preserving the culinary traditions of the Nebrodi. With stricter food-production laws from a unified EU suffocating individual food cultures, Santa Mamma and the Colluras defiantly defend their time-honored practices. The help of both a local shepherd and farmer with deep roots to the territory helps them keep these practices alive. Each morning animals are milked by hand and cheese is made the way it once was: in a copper cauldron set over an open wood fire inside a small stone hut located in a field where the animals graze. When we first spread some of the fresh ricotta onto crusty bread, it was a revelation. The flavor,

redolent of the herbs and grasses eaten by the animals, had a hint of smokiness to it from the fire. It was a far cry from the industrialized mass-produced dairy products sold in supermarkets. The meat from Santa Mamma's free-grazing animals is intensely rich, from their active lives and natural diets. Alfonso can describe the animals' diets as he cooks the meat—oil glistens from the steaks of cattle that grazed on fallen olives. The jet-black pigs native to the park provide a deeply flavored pork and embody the farming philosophies of the Nebrodi. The pigs dwell throughout the hilly terrain, digging their snouts into the dirt, foraging for the acorns, grubs, and roots that make their meat so succulent. Fruits and vegetables are grown organically and in abundance, with the temperate Sicilian climate giving ample bounty to the kitchen. Citrus trees bring fresh oranges to the kitchen, and Giusy slices them up and squeezes their juices over boiled veal shanks to liven up a unique cold salad antipasto. Entire fields of tomato plants ensure busy summers in the kitchen, canning and preserving the fruit for use throughout the year. Santa Mamma's astounding surplus of 12,000 olive trees supply bountiful amounts of oil and—together with the farm's own onions, Alfonso and Giusy's favorite vegetable—form the base for of their cooking. Thinly sliced onions simmer in generous amounts of oil to give a meltingly sweet touch to pasta sauces. Two staples are wild fennel and broccolini, both slow-cooked in pureed tomatoes. Balls of ricotta and coconut rolled in cocoa then refrigerated are delectable, and Sicilian pistachios also make their way into many of Giusy's cakes and sweets. Santa Mamma offers an unparalleled chance to witness a slice of Old World Sicilian country life. If you're seeking an authentic taste of Sicily's rural interior culture, journey into the depths of the Nebrodi Park and enjoy a few farm-fresh meals at Alfonso and Giusy's agriturismo Santa Mamma.

Wildflowers down to the sea

The Santa Mamma Salad

INSALATA SANTA MAMMA

This cold salad, brightened with oranges, offers a vinegary kick and a fresh counterpoint to tender veal. Juicy and tart, it's a unique combination of ingredients that makes for an interesting starter or light lunch.

For the veal:

1 yellow onion, quartered

1 carrot, peeled and chopped

1 stalk celery, chopped

3 veal shanks (6–8 oz./170–226 g each)

For the salad:

1 carrot, peeled and cut into small dice

1 stalk celery, cut into small dice

1 green bell pepper, seeded and cut into small dice

4 caper berries, chopped (about ¼ cup)

1 tablespoon (15 ml) red wine vinegar

2 tablespoons (30 ml) extra-virgin olive oil

Kosher salt and freshly ground black pepper

2 medium navel oranges

1. **Make the veal:** Put the onion, carrot, and celery in a 6-quart pot with 2½ quarts of water and bring to a boil over medium-high heat. Add the veal shanks, reduce the heat to maintain a gentle simmer, and cook until the shanks are fork-tender, 1½ to 2 hours. Remove the veal from the broth and let cool. Refrigerate until well chilled—at least 4 hours and up to overnight. Discard the broth or strain the broth through a fine-mesh sieve and reserve for another use.

2. **Make the salad:** Cut the veal into small, bite-size pieces and transfer to a large bowl. Add the carrot, celery, bell pepper, capers, red wine vinegar, and oil. Season to taste with salt and pepper. Refrigerate for 30 minutes to let the flavors meld.

3. Trim away the peel from the oranges with a pairing knife. Cut the orange segments free from the membrane, and then cut the segments into pieces. Transfer the segments to a bowl, then squeeze the juice from the membranes into the bowl.

4. Fold the oranges and their juices into the cold veal salad and adjust the seasoning if necessary. Serve immediately.

SERVES 4

Free-range pigs of the Nebrodi

NEBRODI

The enchanted landscape of the Nebrodi regional park has eluded tourists' radar, due in part to perhaps the perilous potholed dirt road that runs through it. Those brave enough to journey up the narrow mountain road are rewarded with a haven of green rolling hills, forests thick with brush and Mediterranean vegetation, and red rock outcroppings cascading down to a raging river that flows out to the sea. The Nebrodi speaks to those seeking to get away from the bustling coastal villages and larger cities of the island, to an undiscovered Sicily. Here farms like Santa Mamma continue to work the land as they have for centuries, allowing their animals to live freely in woods and open pastures. The meat offers testament to the character of this place, where humankind has integrated into its landscape but allowed Mother Nature to provide an optimal environment for farming. The park's designation as a protected area ensures that this sanctuary will remain uncontaminated for years to come, and the animals and farmers that call the Nebrodi home will continue to live as they always have in one of Sicily's most pristine locations.

Pasta with Sardines in the Sea
PASTA CON LE SARDE

In Palermo fried sardines are added to this dish full of typical Sicilian flavors. Santa Mamma's isolated mountain location makes sourcing fish a difficult task, and they have aptly named this sauce Pasta with Sardines in the Sea. Wild fennel grows all over Sicily. It differs from the variety of central Italy, resembling the feathery green fronds that grow from the top of our fennel bulbs back home. Look for the fennel bulbs that have the greatest number of fronds to re-create this classic recipe, with or without the sardines.

2 medium bulbs fennel, fronds attached

¼ cup (59 ml) extra-virgin olive oil

2 small yellow onions, thinly sliced

Kosher salt and freshly ground black pepper

2 tablespoons tomato paste

3 tablespoons pine nuts

2 tablespoons currants

1 28-ounce (793 g) can plum tomatoes, crushed by hand

1 pound (453 g) bucatini

Grated Parmigiano Reggiano or pecorino cheese, for serving

1. Remove the fennel fronds from the fennel and roughly chop them. Set aside in a bowl. Cut the fennel bulbs in half lengthwise and remove the core. Thinly slice the fennel halves, preferably on a mandoline or with a very sharp knife.

2. Bring a large pot of well-salted water to a boil over high heat. Add the sliced fennel and blanch for 2 to 3 minutes. Remove the fennel with a fine-mesh sieve and immediately run under cold water. Reserve the boiling water to cook the pasta in. Squeeze the excess water from the fennel over a small bowl and reserve the water.

3. Put the oil, onions, ¼ teaspoon of salt, and a few grinds of black pepper in a 4-quart pot over low heat. Cook the onions, stirring occasionally, until they are meltingly tender and very lightly golden, 30 to 40 minutes. Raise the heat to medium and add the tomato paste; stir well to combine, and cook until the tomato paste darkens, about 5 minutes. Add the fennel, fennel fronds, reserved water, pine nuts, currants, and a pinch of salt and stir well to combine. Add the tomatoes and bring to a simmer. Reduce the heat to maintain a gentle simmer and cook until the flavors have melded together and the sauce has thickened, about 1 hour. If the sauce becomes too thick, add a bit of water as necessary.

4. Cook the bucatini in the same large pot of boiling water you used to blanch the fennel fronds until al dente, according to the package instructions. Reserve ¼ cup of the pasta water and then drain the pasta well. Toss the bucatini with the sauce, adding some of the reserved pasta water if needed. Serve in a large bowl with the grating cheese on the side.

SERVES 6

Wild fennel

Rotelle Pasta with Broccolini Sauce

ROTELLE AI BROCCOLETTI

Thinly chopped broccolini melts into this long-simmered tomato sauce full of sweet Sicilian flavor.

Kosher salt

2 bunches broccolini (about 8 oz. /226 g each), trimmed

¼ cup (59 ml) extra-virgin olive oil

1 medium yellow onion, thinly sliced

2 small cloves garlic, thinly sliced

1 28-ounce (793 g) can crushed tomatoes

1 pound (453 g) rotelle pasta

1. Bring a large pot of well-salted water to a boil. Add the broccolini and cook until just tender, 2 to 3 minutes. Remove the broccolini from the water with tongs to a large colander, run under cold water, drain, and then let dry on kitchen towels. Chop the broccolini into ¼-inch pieces and set aside. Reserve the boiling water to cook the pasta.

2. Heat the oil in a 10-inch straight-sided skillet over medium-low heat. Add the onion, garlic, and a generous pinch of salt; cook until very tender and lightly golden, 15 to 20 minutes. Add the crushed tomatoes and broccolini and cook until the broccolini is very tender and the sauce is flavorful, about 1 hour. Season to taste with salt.

3. Cook the pasta in the pot of reserved water until al dente, according to package directions. Drain the pasta and toss with the sauce. Serve immediately.

SERVES 6

(clockwise) Preparing ricotta cheese, Heating goat's milk in copper cauldron, Steaming goat's-milk ricotta

Ground Veal and Aged Caciocavallo Filled Veal Rolls

INVOLTINI DI VITELLO

2 small yellow onions

¼ cup (59 ml) extra-virgin olive oil

Kosher salt and freshly ground pepper

¾ pound (340 g) ground veal

¼–½ cup (59–118 ml) dry white wine

2 tablespoons tomato paste

¾ cup fine bread crumbs

½ cup grated aged caciocavallo cheese or Parmigiano Reggiano

½ cup shredded Toma cheese

¼ cup chopped flat-leaf parsley

10 2-ounce (60 g) veal cutlets, ⅛ inch thick

12 bay leaves

Laurel trees grow profusely throughout southern Italy, and bay leaves are a mainstay in each region. Here the herbal and floral qualities of the leaves are absorbed into the veal rolls as they bake, infusing the entire dish with subtle flavor. If there is any leftover filling after making the involtini, use it to make a pasta sauce.

1. Cut the onions in half lengthwise and then cut three of the halves into thin slices. Cut the remaining onion half into quarters lengthwise; separate the pieces and set aside in a small bowl.

2. Heat 3½ tablespoons of the oil in a 12-inch skillet over medium-low heat. Add the sliced onions, a pinch of salt, and a few grinds of pepper; cook, stirring often, until very tender and golden brown, 15 to 20 minutes. Raise the heat to medium-high and add the ground veal and season with salt. Cook, breaking up the clumps of meat with a wooden spoon, until browned, 5 to 6 minutes. Add the white wine and cook until reduced by half, 2 to 3 minutes. Stir in the tomato paste and cook until it has darkened in color, 2 to 3 minutes. Stir in the bread crumbs to absorb most of the liquid, remove from the heat, and set aside to cool. Once cooled, stir in the caciocavallo, Toma, and parsley. Season to taste with salt and pepper.

3. Position a rack in the center of the oven and heat the oven to 425°F (220°C).

4. Place one of the veal cutlets on a cutting board and season with salt and pepper. Place 2 tablespoons of filling in the center of the cutlet and then roll the veal up. Set aside on a large plate and continue with the remaining cutlets.

Goats of the Nebrodi

5. Arrange the veal rolls in an 8 x 8-inch baking dish. In between each roll, set a bay leaf and a onion wedge. Drizzle with the remaining olive oil and bake in the oven until cooked through and lightly golden brown, 30 to 33 minutes. Transfer to a platter and drizzle with any of pan juices. Serve immediately.

SERVES 5

TENUTA ROCCADIA

The sprawling Tenuta Roccadia agriturismo, a complex of low-lying stucco buildings painted in brilliant reds, yellows, and oranges, is tucked into a hill and nestled among groves of citrus, olive, and almond trees. Here Pietro Vacirca has found his personal paradise, managing Tenuta Roccadia and catering to the thousands of visitors who make their way to the sun-kissed property. Noted as the very first agriturismo to open in Sicily over twenty years ago, Tenuta Roccadia has aged gracefully over the years, developing into a unique vacation resort. Today the complex has grown to include two large swimming pools, an archery range, soccer field, walking trails that ramble through orange and lemon trees, past horses, cows, peacocks, and pigs, and two large restaurant halls that showcase Sicilian feasts. Families come to spend their vacation to relax and enjoy the rhythms and pace of a working Sicilian farm.

The agriturismo sits in an unparalleled geographic position, sheltered from wind and with very little rainfall, and has an ideal environment for the citrus to thrive. Throughout the winter months, trees around the estate are heavy with lemons, mandarins, clementines, and oranges. Plucking the fruit straight from the branches reveals an explosion for the senses—a true taste of Sicily, leaving fingers sticky and hands perfumed with citrus for the rest of the day. While Sicily's impoverished past led to massive migrations around the world, oranges and lemons have brought wealth and prosperity to the southeast corner of the island, and Tenuta Roccadia has reaped the benefits of its strategic location.

Tenuta Roccadia has amassed a loyal following of Sicilians who make weekly Sunday pilgrimages. In open-air banquet rooms, large glass windows overlook grazing animals in verdant pastures that extend down to the sea, and long wooden tables are joined together for families and friends to sit together and enjoy marathon lunches that require stamina, strength, and a hearty appetite. These celebratory feasts have made the agriturismo famous. Seasonal menus focus on the farm's own products. Citrus plays a significant role in the kitchen: Pork shanks are braised fork-tender in a mildly acidic citrus-infused white wine sauce, while grilled wedges of the farm's own provolone are sprinkled with pistachios and drizzled with a blood orange and honey reduction. An orange Jell-O-type sweet has become a mainstay on the menu, as has a Sicilian-style after-dinner shot made from peeled citrus fruits (lemons, mandarins, or oranges) infused in alcohol for weeks. Citrus marmalade made from a family recipe accompanies slices of crusty bread or wedges of sharp cheese. Almonds give a crunchy bite to eggplant caponata and peperonata, and are pureed with basil for a classic Sicilian pesto. The southern Italian favorite horse meat also makes its way onto the menu, often grilled and bathed in the farm's own olive oil. The genuine cuisine from its own home-grown products has brought Tenuta Roccadia impressive popularity.

Orange grove

Grilled Provolone with Blood Orange Reduction

PROVOLA ALLA GRIGLIA CON RIDUZIONE DELL'ARANCIA ROSSA

½ cup (118 ml) freshly squeezed blood orange juice

1 tablespoon mild honey

Kosher salt and freshly ground black pepper

½ tablespoon (7 ml) extra-virgin olive oil

8 ½-inch-thick wedges mild provolone cheese

2 tablespoons chopped pistachio nuts

Pistachio nuts and a concentrated blood orange reduction drizzled over griddled cheese make for a perfect marriage of sweet melting flavors with a crunchy texture. Any firm cow's-milk cheese should work well in this recipe; just be sure that it's not so sharp that it overpowers the delicate sauce.

1. Put the orange juice and honey in a small saucepan and bring to a boil over medium-high heat, stirring until the honey dissolves. Cook until the mixture is reduced by half and is thickened and slightly syrupy, 8 to 12 minutes. Remove from the heat and season with a pinch of salt and a few grindings of black pepper.

2. Heat a grill pan or griddle over medium-high heat until smoking. Rub the pan with the olive oil and then grill the cheese until lightly golden and soft, 30 seconds to 1 minute per side.

3. Transfer two wedges of cheese to individual plates. Drizzle with the orange reduction and sprinkle with the chopped pistachio nuts. Serve immediately.

SERVES 4

Grilled Meatballs with Bay Leaves and Lemon

POLPETTE ALLA GRIGLIA CON ALLORO E LIMONE

At Tenuta Roccadia these meatballs are grilled between leaves of fresh-picked leaves from the citrus trees that surround the property. Fresh bay leaves work just as well, and we added slices of lemon to replicate the farm's dish. The citrus and laurel flavor penetrates into the meat as it cooks over the hot fire for a tasty and different way of cooking meatballs.

3/4 cup bread crumbs

1 pound (453 g) ground beef

3 large eggs

1/3 cup grated Parmigiano Reggiano

1/4 cup chopped flat-leaf parsley

Kosher salt and freshly ground
  black pepper

Extra-virgin olive oil, as needed

24 bay leaves, preferably fresh

16 thinly sliced lemon rounds (from
  2 small)

1. Prepare a medium gas or charcoal grill fire.

2. In a small bowl, dampen the bread crumbs with 2 tablespoons of water.

3. In a large bowl, mix together the bread crumbs, ground beef, eggs, Parmigiano, parsley, 1½ teaspoons of salt, and a few grinds of pepper. Mix the ingredients well until the mixture is homogeneous.

4. Shape the meatballs: Lightly dampen your hands with the oil and then gently scoop up a handful of meat (about ¼ cup) and roll it into a nice even ball. Transfer to a large plate and continue forming the meatballs, dampening your hands as necessary to keep the mixture from sticking to your hands.

5. Thread six bay leaves, four lemon rounds, and three meatballs onto each metal skewer (you will need four skewers) in an alternating pattern. Oil the grill grate and then arrange the skewers on the grill. Grill until nicely browned on one side, 5 to 6 minutes. Gently turn the skewers over and cook until golden brown on the other side and cooked through, 5 to 6 minutes more. Serve immediately.

MAKES 12 MEATBALLS

Sicilian Eggplant and Peppers

CAPONATA

1/4 cup (59 ml) extra-virgin olive oil

1 medium red bell pepper, cut
   into 1/4-inch strips

Kosher salt

1 medium eggplant, cut into
   1/2-inch cubes

1 medium red onion, cut into
   medium dice

1 carrot, peeled and cut into
   medium dice

1 stalk celery, cut into medium dice

1/3 cup chopped almonds

1/3 cup pitted green olives, such
   as picholine, smashed

2 tablespoons capers

1 cup crushed tomatoes

2 tablespoons granulated sugar

3 tablespoons (44 ml) white wine
   vinegar

1/4 cup packed torn basil leaves

Freshly ground black pepper

Caponata is inescapable if you are visiting Sicily. Every family has their own version of this dish, which can include everything from cocoa and cinnamon to nuts and raisins. At Tenuta Roccadia they add chopped almonds that they harvest from their land to add a bit of crunchy texture. This really tastes great the day after making it, after the flavors have had time to come together.

1. Heat 1 tablespoon of the olive oil in a 12-inch skillet over medium heat. Add the pepper and a pinch of salt and cook, stirring occasionally, until tender and beginning to brown, 5 to 7 minutes. Transfer to a clean plate. Add another 2 tablespoons of oil to the skillet along with the eggplant and a pinch of salt and cook, stirring often, until the eggplant is browned all over and tender, 5 to 7 minutes. Transfer the eggplant to the plate with the peppers.

2. Add the remaining 1 tablespoon of oil to the skillet along with the onion, carrot, celery, and a pinch of salt. Cook until tender and lightly browned, 5 to 7 minutes. Add the almonds, olives, and capers, and cook for 2 minutes. Stir in the crushed tomatoes, sugar, and vinegar, and bring to a simmer. Add the peppers and eggplant to the pan and cook until the flavors meld together, about 5 minutes. Stir in the basil leaves and season to taste with salt and pepper. Transfer to a platter and serve hot or at room temperature.

SERVES 6

Caponata

Pistachio Semifreddo

SEMIFREDDO AL PISTACCHIO

Pistachio trees thrive in eastern Sicily in the volcanic soil surrounding Mount Etna. The nut has become a key component in the many traditional sweets of the island.

1. Put the pistachios in a food processor and pulse until finely chopped.

2. In the pan of a double boiler, combine the egg yolks, ¾ cup of the sugar, and the nuts. Place the pan over barely simmering water and whisk constantly until the mixture is thickened and doubled in volume and its temperature reaches 175°F (80°C) on a candy thermometer, about 4 minutes. Remove the pan from the heat and immediately submerge in an ice-water bath, stirring constantly to bring down the temperature.

3. Meanwhile, in a stand mixer fitted with the whisk attachment, beat the heavy cream on medium speed until soft peaks form. Gradually add the remaining 2 tablespoons of sugar in a steady stream, and continue to beat until stiff peaks form. Transfer to a large bowl.

4. In a stand mixer fitted with the paddle attachment, beat the egg yolk mixture until it becomes thick and pale in color, 3 to 5 minutes.

5. Line an 8½ × 4½-inch loaf pan with plastic wrap. Using a spatula, add a quarter of the whipped cream to the egg mixture and stir together gently to lighten the base. Fold the remaining whipped cream into the egg mixture and then spoon the custard into the prepared loaf pan. Cover with plastic wrap and freeze until firm, about 4 hours.

6. To serve, remove the semifreddo from the freezer and let stand at room temperature for 3 to 4 minutes to soften slightly. Invert the semifreddo onto a cutting board, remove the plastic wrap, and slice the semifreddo. Transfer the slices to individual plates and serve.

SERVES 8

¼ cup unsalted shelled pistachio nuts

10 large egg yolks

¾ cup plus 2 tablespoons granulated sugar

1½ cups (360 ml) heavy cream

Blood Oranges

The area around the small town of Lentini in southwestern Sicily has a unique climate due to its close proximity to North Africa. Here sunshine prevails with very little rain, affording extended growing seasons. This has blessed the area with fruitful citrus trees that bear the prized blood-red orange. Considered the hallmark of Sicilian citrus, the blood oranges of Lentini are known as Morro, and believed to have been developed from a strand of citrus brought over to Sicily by the Arabs in the ninth or tenth centuries. They possess an orange flesh speckled with ruby red; it's juicy and sweet with a hint of acid. The northern Italian soft drink manufacturer San Pellegrino was keen on the sweet citrus taste of blood oranges, and developed a soda, Aranciata Rossa, that includes a percentage of real juice from Lentini oranges and offers a refreshing sip of sunny Sicily.

Orange Gelatin
GELO D'ARANCIA

This creamy, luscious dessert is a classic in the Sicilian repertoire. Pureed melon, lemon or tangerine juice are also used as a base to flavor this dairy-free custard, taking advantage of the islands fruitful bounty.

1. Whisk together the sugar and cornstarch in a 4-quart saucepan. Gradually whisk in the orange juice and vanilla, making sure there are no lumps. Bring to a boil over medium heat, stirring constantly. Continue to boil, stirring constantly, until the mixture thickens and begin to sputter, about 2 minutes.

2. Distribute the pistachio nuts among eight 4-ounce ramekins. Pour the orange mixture into each ramekin, filling to about ¼ inch from the top. Refrigerate the mixture until it solidifies — at least 2 hours and up to 24 — before serving.

3. To serve, unmold the gelo by running a butter knife around the rim of the ramekin, dip the bottom of the ramekin into hot water, and then invert the ramekin onto an individual plate. Complete with the remaining ramekins.

1 cup granulated sugar

½ cup cornstarch

4 cups (about 1 liter) freshly squeezed orange juice (from about 8 large oranges), strained

½ teaspoon vanilla extract

¼ cup chopped pistachio nuts

Gelo d'Arancia

BAGLIO FONTANA

Nicola Di Vita holds a deep passion for the gastronomic wonders surrounding his homeland of Trapani in the northwest corner Sicily. He discovered the pleasures of Sicilian cuisine at a young age, through visits to his grandfather's farm during the grape harvest. His fondest memories are at the dining room table, where he would join the farmers for their midmorning lunch and then happily sit through a second feast with his relatives. Today he jokes about the size of his belly, a testament to the same voracious appetite he once had, without the same metabolism.

A descendant of a noble family from Messina, Nicola has taken active steps to preserve Trapanese food culture as well as his family's legacy. In the early 1800s, after Palermo succeeded Messina as the capital of Sicily, his family relocated to the west coast to establish roots in the newly formed kingdom. There, in the countryside surrounding Trapani, they bought land and built several farmhouses known as bagli. Each baglio specialized in different agricultural activities, including the production of wine, grains, and olive oil, and the rearing of animals. Trapani's coastal location also boasted the island's largest fishing fleets, and the family established a tuna cannery called a tonnara. Boats would dock and unload their catches, and the fish were wheeled by mule-driven carts to the tonnara, where they were filleted, cooked, and canned.

Over the years, the economic landscape of Sicily has changed, with farms abandoned and fields neglected. The once thriving tuna population has been decimated by overfishing, too, leaving the processing plants ghostly empty. Nicola has sought to revitalize his inheritance by restoring the brick tuna plant into a boutique waterfront hotel, and one of the farmhouses into a country agriturismo. This new "tourism" direction has breathed life into his family's businesses. The Di Vitas now join a growing list of Sicilians catering to the many vacationers visiting Sicily each year seeking alternative types of accommodations. A visit to either the tuna plant or the farm assures an unforgettable holiday, where plenty of good eating offers insight into traditional Sicilian culture, kept alive and preserved by an always hungry Nicola.

The restored and revitalized Baglio Fontana agriturismo offers a glimpse into Sicilian farm life. A square structure built around an open-air piazza served as a sustainable miniature village that housed the families who worked the land for the Fontana family. Surrounded by vineyards, the baglio was once an active cantina. Nicola has left all of the original equipment in place. An enormous wooden press greets visitors to the restaurant. Behind the bar sits a large stone sink, where the women of the baglio would stain their toes, stomping the grapes of the fall harvest. The juices would flow directly into a fixed basin; pumps would then transfer the liquid to wooden barrels, which now line the dining room. Nicola offers a dessert wine made in part from vintages dating as far back as the mid-1800s, bringing to life a taste of the golden years of the farm's winemaking tradition.

Today the farm has diversified its production to include olive oil and vegetables in addition to wine. With the waters of the Tyrrhenian Sea lapping at the shore a few miles away, fish is a staple at Baglio Fontana. The photographs decorating the dining room walls speak of Nicola's love for his native waters, and highlight two of Trapani's most prized ingredients, fish and salt. A platter of cured fish begins an all-seafood feast: thin slices of lightly smoked swordfish, slightly chewy air-dried tuna, and shavings of cured salty roe sacs, known as bottarga, from female tuna. First courses display Arabic influences on Trapanese cuisine with couscous and tabouli. One of the area's most famous dishes, seafood couscous, is a laborious effort that involves steaming the grains over a tomato-infused fish broth, then mixing them by hand until the couscous has achieved the proper texture. It's served with shellfish, fried calamari, and a fillet of steamed fish, generously garnished with almonds. Eating the dish awakens the palate. Local spiny lobsters make an incredibly intense broth in which broken bits of spaghetti are cooked to make for a delicious soup whose taste is alone worth traveling to the western end of the island to experience. Tuna meatballs cooked in a sweet-and-sour agrodolce sauce utilize the farm's own vinegar to impart an acidic, tingling bite. Although this is a working farm surrounded by fertile hills, vineyards, and olive trees, a meal at Baglio Fontana pays its respects to the island's coastlines and waters. Fruits of the sea have blessed the Di Vita table, bestowing success and full bellies to Nicola and his family—a tradition he now proudly passes on to his guests.

Baglio Fontana courtyard

Tabouli with Green Olives, Sun-Dried Tomatoes, and Grilled Zucchini

TABBULEH CON OLIVE VERDI, POMODORI SECCHI, E ZUCCHINE

Sicily's Middle Eastern roots are perhaps most apparent in some of the region's dishes. Tabouli, a common first course eaten around the island, has roots that can be traced back to Arabic influences in the 900s. This first course or side dish is loaded with fresh flavor and offers a glimpse into a seemingly non-Italian but very authentic Sicilian ingredient.

Kosher salt

1 cup fine bulgur wheat

1 small zucchini, cut lengthwise into ½-inch strips

3 tablespoons (44 ml) extra-virgin olive oil

Freshly ground black pepper

½ cup small green olives, pitted and chopped

½ cup chopped sun-dried tomatoes (not in oil)

¼ cup loosely packed basil leaves, thinly sliced

1 tablespoon chopped mint

1 tablespoon lemon zest

1 tablespoon (15 ml) lemon juice

¼ cup fine diced caciocavallo or sharp provolone cheese

1. In a small saucepan, bring 1 cup of water with ½ teaspoon of salt to a boil over high heat. Put the bulgur in a medium bowl, add the boiling water, cover with a kitchen towel, and let it stand until just tender, about 15 minutes.

2. Prepare a medium-high gas or charcoal grill. Brush the zucchini slices all over with ½ tablespoon of the oil and season with salt and pepper. Grill on both sides until golden brown and tender, about 2 minutes per side. Let cool and then chop into pieces.

3. In a large bowl, combine the wheat, zucchini, olives, sun-dried tomatoes, basil, mint, and lemon zest, and stir well to combine. Stir in the lemon juice and the remaining oil. Gently stir in the cheese and season to taste with salt and pepper. Serve immediately or refrigerate until ready to serve.

SERVES 4–5

Fish Broth

BRODO DI PESCE

This broth acts as the base for several of Baglio Fontana's seafood dishes, enhancing their overall depth of flavor. Called for in both the lobster pasta and fish couscous, it is the main reason behind each recipe's great taste.

1. Cut the bones into 3- to 4-inch pieces. Rinse the bones under cold water and then set in a container. Cover the bones with cold water and soak in the refrigerator, changing the water several times, until the water becomes clear, at least 8 hours and up to overnight.

2. Heat the oil in a large stockpot over medium heat. Add the garlic, onion, and carrot, and cook, stirring occasionally, until tender but not browned, 5 to 7 minutes. Add the tomato paste to the pot and stir to coat the vegetables; cook for 2 minutes. Pour in the white wine and simmer until the alcohol burns off, about 5 minutes.

3. Drain the fish bones and add them to the pot. Cover with a lid and steam the bones until they are opaque, 4 to 6 minutes. Add 3 quarts of water, enough to cover the bones, along with the bay leaves, parsley, and thyme sprigs. Slowly bring up to a simmer, skimming away any scum that floats to the surface. Cook the broth at a gentle simmer until it's fragrant, 30 to 45 minutes. Remove from the heat and let cool slightly. Strain the broth by ladling it through a fine-mesh sieve lined with dampened cheesecloth into a 6-quart pot.

MAKES 2 1/2 QUARTS

5 pounds (2.2 kg) fish bones, such as halibut, bass, or flounder, tails, heads, fins, and any skin removed

1 tablespoon (15 ml) extra-virgin olive oil

6 cloves garlic, smashed

1 medium yellow onion, sliced

1 carrot, peeled and chopped

¼ cup tomato paste

1 cup (240 ml) dry white wine

2 bay leaves

2 sprigs parsley

1 sprig thyme

Fish Couscous
CÙSCUSU DI PESCE

1 recipe fish broth (previous page)

Kosher salt and freshly ground black pepper

½ cup finely chopped blanched almonds

5 tablespoons chopped flat-leaf parsley, plus more for garnish

1 yellow onion, finely chopped

1 tablespoon minced garlic

¼ cup (59 ml) extra-virgin olive oil

2 cups couscous

3 bay leaves

1 cinnamon stick

2 pounds (900 g) mussels, scrubbed and debearded

Pinch of crushed red pepper flakes

1 cup (240 ml) dry white wine

6 4-ounce (115 g) halibut fillets

Locally caught fish paired with tiny yellow pearls of semolina flour has become recognized as the classic Trapanese dish, reminiscent of the city's past under Arab rule. An essential step in making proper couscous is ladling fish broth into the grain and rubbing it together by hand to break up any clumps and to separate the granules. It is then set in a colander over simmering fish broth, which both steams and infuses the couscous. Although this recipe may seem daunting to the non-Trapanese-native, it actually comes together quickly once the essential broth has been made.

1. In a large bowl, mix together 1 cup hot water, 1 cup fish broth, 1 teaspoon salt, and ½ teaspoon pepper. Stir in ¼ cup of the almonds, 1 tablespoon of the parsley, 1 tablespoon of the onion, and 1 teaspoon of the garlic.

2. Put 2 tablespoons of the oil in a large bowl and add half of the couscous. Ladle in some of the fish broth and sprinkle with 1 tablespoon of parsley. Rub the couscous back and forth between your hands to separate the granules. Add the remaining couscous to the bowl. Ladle in some more of the fish broth and sprinkle with 1 tablespoon of parsley. Rub the mixture back and forth between your hands. Keep rubbing the couscous until the granules swell. Add 2 tablespoons of the almonds and mix well to combine.

3. Transfer the couscous to a metal colander. Add the bay leaves, cinnamon stick, the remaining almonds, the remaining onion, 1 teaspoon of garlic, and ½ teaspoon black pepper.

4. Pour the remaining fish broth into a 6-quart pot and line with foil around the rim. Set the colander over the pot and adjust the foil as necessary so the colander sits tightly in the pot. Bring the fish broth to a simmer over medium heat and steam the couscous, without stirring or a lid, for

10 minutes. Give the couscous a stir and continue cooking until tender and fragrant, 10 to 12 minutes more. Transfer to a large bowl and discard the cinnamon and bay leaf. Add 1 to 2 ladlefuls of Fish Broth to the bowl, stir to combine, cover with a kitchen towel, and let rest 15 minutes.

5. In a straight-sided 12-inch skillet, heat the remaining 2 tablespoons of oil over medium-high heat. Add the remaining 1 teaspoon of garlic and cook until lightly golden, about 1 minute. Add the mussels and pepper flakes, shaking the pan to coat the mussels with oil. Add the white wine and bring to a boil. Stir in the remaining parsley, cover the pan, and reduce the temperature to maintain a steady simmer. Cook until the mussels open up, 5 to 8 minutes. Remove from the heat.

6. Gently poach the halibut in the remaining fish broth in the pot over medium heat until the fish is opaque and flaky, 6 to 10 minutes.

7. To serve, spoon the couscous into individual soup bowls. Drizzle each with some of the liquid from the mussels. Top each bowl with the mussels and a piece of halibut. Garnish with some chopped parsley.

8. Transfer the remaining Fish Broth to a tureen and serve on the side. Serve immediately.

SERVES 6

Spaghetti in Lobster Broth
PASTA IN BRODO CON ARAGOSTA

The clawless lobsters taken from Sicilian waters are slightly sweeter than the cold-water crustaceans of North America. This insanely delicious recipe can be made with either type of lobster and offers deeply layered flavors that resonate with the delicate taste of the shellfish. Although intricate and involved, this dish will deliver praise of the highest kind from whoever is lucky enough to experience it. Slowly simmering broken bits of spaghetti in the lobster and fish broth infuses even more lobster taste into every bite taken from this Sicilian delicacy.

2 whole live lobsters (about 1¼ pounds/550 g each)

2 tablespoons (30 ml) extra-virgin olive oil, plus more for serving

1 medium yellow onion, cut into fine dice

1 clove garlic, thinly sliced

2 tablespoons tomato paste

2 quarts (2 liters) fish broth (page 307)

Kosher salt

Pinch of crushed red pepper flakes

1 tablespoon chopped flat-leaf parsley, plus more for garnish

1 pound (453 g) spaghetti, broken into 2-inch pieces

Freshly ground black pepper

1. Lay the lobsters on their backs on a cutting board. Position the tip of a chef's knife just below the large claws. In one swift motion, insert the knife into the body and chop down through the head. Split the lobsters in half lengthwise, clean, and then cut into pieces.

2. Heat the oil in an 8-quart saucepan over medium heat. Add the onion and garlic and cook until tender and translucent, 4 to 5 minutes. Add the tomato paste, and cook until fragrant, about 1 minute. Add the lobster pieces, cover, and cook until the shells begin to turn light red, 3 to 4 minutes. Add the Fish Broth, bring up to a boil, reduce the heat to maintain a very gentle simmer, and cook until the lobster is just cooked through, 2 to 4 minutes.

3. Remove the meat from the tails, claws, and knuckles and set aside in a bowl. When it's cool enough to handle, cut the lobster meat into small pieces and set aside.

Trapanese sea salt

4. Return the shells to the pan along with a pinch of salt, crushed red pepper flakes, and the parsley. Continue to simmer until the soup has thickened some and is very flavorful, about 45 minutes. Strain the soup through a fine-mesh sieve into a clean 8-quart pot. Return the pot to medium heat and bring the broth to a simmer. Add the spaghetti to the pan and cook, stirring occasionally, until the pasta is just al dente, about 9 minutes. Add the reserved lobster meat to the pan and cook until just warmed through. If the soup is too dense, thin with a bit more fish broth or a little warm water. Season to taste with salt and pepper.

5. Ladle the soup into shallow bowls, garnish with some chopped parsley and a drizzle of oil, and serve.

SERVES 6

Sweet-and-Sour Swordfish with Cipolline Onions

PESCE SPADA CON CIPOLLINE IN AGRODOLCE

The contrasting flavors of sweet sugar and sour vinegar are widespread through-out Sicilian cooking and referred to as agrodolce. At Baglio Fontana, chef Salvatore simmers meaty fillets of locally caught swordfish in the tangy solution. He adds his own spin by using two types of vinegar: white wine for its assertive kick and balsamic for a nice hint of sweetness.

3 tablespoons (44 ml) extra-virgin olive oil

16 small cipolline onions, halved lengthwise

1 yellow onion, halved and thinly sliced

Kosher salt

1 cup (240 ml) white wine vinegar

2 tablespoons granulated sugar

1 cup (240 ml) balsamic vinegar

4 6-ounce (170 g) swordfish steaks (each about ½ inch thick)

Freshly ground black pepper

1 teaspoon chopped thyme leaves

1. Heat 2 tablespoons of the oil in a 12-inch straight-sided sauté pan over medium-high heat. Add the cipolline onions, sliced onion, and a pinch of salt; cook until just tender and translucent, 5 to 7 minutes. Add 1 cup of water, the white wine vinegar, and the sugar to the pan. Bring to a boil, reduce the heat, and simmer for 5 minutes. Add the balsamic vinegar and simmer until the onions are very tender and the liquid has reduced by a quarter, about 10 minutes.

2. Season the swordfish with salt and pepper. Heat the remaining 1 table-spoon of oil in a 12-inch skillet over medium-high heat. Add the swordfish and cook until golden brown, 1 to 2 minutes. Flip the fish over carefully with a spatula and cook until the other side is golden brown, about 2 minutes. Carefully transfer the swordfish to the pan with the onions. Press a piece of foil or parchment paper down on the fish, and then cover with a lid. Cook the fish over medium-low heat at a very gentle simmer until it's just cooked through, 10 to 12 minutes. Remove the pan from the heat and let rest for 5 minutes.

3. Transfer the swordfish steaks to individual plates. Stir the thyme into the onion sauce and season to taste with salt and pepper. Divide the onion mixture over the swordfish steaks and serve.

SERVES 4

Cassata Deconstructed

CASSATA

Perhaps Sicily's most famous dessert, Cassata Siciliana's various ingredients respect the island's eclectic past as the crossroads and melting pot of some of the world's most important cultures. In the dish's original form, a sponge cake filled with ricotta cream mixed with dark chocolate pieces is encased in an almond marzipan, decorated with a stripe of pistachio cream, and dotted with candied fruit. In this recipe, the dessert combines all of the flavor components of the classic sweet in a modernized, somewhat lighter version.

1. **Make the chocolate mousse:** Melt the chocolate with the butter in a medium heatproof bowl over a pan of barely simmering water. Set aside.

2. In a stand mixer with the whisk attachment, beat the egg whites on medium speed until soft peaks form, 2 to 3 minutes. Gradually pour in the sugar, raise the speed to medium-high, and continue beating until glossy stiff peaks form.

3. Put the heavy cream in a large bowl and beat with a handheld mixer on medium speed until medium-stiff peaks form, 2 to 3 minutes.

4. Whisk a spoonful of the whipped cream into the melted chocolate. With a rubber spatula, fold in the remaining whipped cream and then the egg whites. Transfer to a pastry bag fitted with a star tip and refrigerate.

5. **Make the candied orange peel:** With a sharp paring knife, cut the orange peel away from the flesh into 2-inch-thick strips. Remove almost all of the white pith from the peel (leaving about ⅛ inch of the pith intact). Cut the peel lengthwise into ¼-inch strips (you should have about ½ cup of peel).

(Continued)

For the chocolate mousse:

5 ounces (145 g) bittersweet chocolate, chopped

2 tablespoons (¼ stick) unsalted butter

2 egg whites

¼ cup granulated sugar

½ cup (118 ml) heavy cream

For the candied orange peel:

1 large orange

¾ cup granulated sugar

For the ricotta cream:

12 ounces (340 g) ricotta cheese

2 tablespoons granulated sugar

1 teaspoon lemon zest

1 teaspoon orange zest

To assemble:

1 recipe sponge cake (see "Basic Recipes")

¼ cup (59 ml) light rum

½ cup chopped blanched almonds

½ cup chopped unsalted pistachio nuts

6. Fill a 2-quart saucepan three-quarters of the way with water and bring to a boil over high heat. Add the orange peel and cook to remove some of the bitterness, 4 to 5 minutes. Drain the peel, run under cold water, and drain again.

7. In the same pan, add ½ cup of the sugar and ¼ cup of water; bring to a boil over medium-high heat. Add the orange peel, reduce the heat to maintain a very gentle simmer, and cook until the peel is translucent, 20 to 25 minutes. Drain the peel.

8. Set a rack over a baking sheet lined with parchment paper. Put the remaining ¼ cup of sugar in a bowl. Toss the peel in the sugar, in batches, and then shake in a sieve to remove any excess sugar. Arrange the orange peel on the rack and leave to dry, 3 to 4 hours.

9. **Make the ricotta cream:** In a large bowl, mix together the ricotta, sugar, lemon zest, and orange zest until combined.

10. **To assemble,** cut the sponge cake into ½-inch cubes. In a small bowl, mix the rum with 1 tablespoon of water. Divide the cake among six 4-ounce glasses. Brush the cake cubes with a little of the rum mixture. Pipe the chocolate mousse over the ricotta cake. Divide the ricotta cream among the cups. Sprinkle each cup with the almonds and pistachio nuts and then top with the candied orange peel.

SERVES 6

Modernized Sicilian Cassata

RESOURCES

1. SARDEGNA

Sa Mandra:
Strada Aeroporto Civile Podere 21
Alghero (SS)
79999150
www.aziendasamandra.it

Sa Tiria:
SS 195 KM 76.500
Teulada (CA)
709283704
www.satiria.it

Il Muto di Gallura:
Loc.Fraiga
Aggius (OT)
79620559
www.mutodigallura.com

2. TOSCANA

Gualdo del Re:
Località Notri 77
Suvereto (LI)
565829888
www.gualdodelre.it

Terra Etrusca:
Via Piemonte, 6
Capalbio Scalo (GR)
564898699
www.terraetrusca.it

Fattoria San Martino:
Via Martiena, 3
Montepulciano (SI)
578717463
www.fattoriasanmartino.it

3. UMBRIA

Villa Dama:
Frazione Torre dell'Olmo
Gubbio (PG)
075 9256130
www.villadama.it

Malvarina:
Località Malvarina 32
Assisi (PG)
758064280
www.malvarina.it

Fonte Antica:
fraz.Campi
Norcia (PG)
743828523
www.fonteantica.it

4. LE MARCHE

Costa della Figura:
Strada Costa della Figura 30
Montefelcino (PU)
721729428
www.costadellafigura.com

Il Giardino degli Ulivi:
Località Castelsantangelo
Castelraimondo (MC)
3383056098
www.ilgiardinodegliulivi.com

5. LAZIO

Casale Verdeluna:
Strada Vicinale della Civitella
Piglio (FR)
775503051
www.casaleverdeluna.it

Le Mole sul Farfa:
Strada delle Mole
Mompeo (RI)
3284603412
www.fiumefarfa.eu

6. ABRUZZO

Le Magnolie:
Contrada Fiorano 83
Loreto Aprutino (PE)
858289534
www.lemagnolie.com

Campoletizia:
Contrada Elcine
Miglianico (CH)
871951225
www.campoletizia.it

Pietrantica:
Contrada Decontra 21
Caramanico Terme (PE)
085922188
www.agripietrantica.com

7. MOLISE

Masseria Santa Lucia:
Contrada Santa Lucia 23
Agnone (IS)
865779846
www.masseriasantalucia.com

I Dolci Grappoli:
Contrada Monte Altino 23/a
Larino (CB)
874822320
www.dolcigrappoli.it

8. CAMPANIA

Il Giardino di Vigliano:
Via Vigliano 3
Massa Lubrense (NA)
815339823
www.vigliano.org

Il Cortile:
Via Roma 43
Cicciano (NA)
818248897
www.agriturismoilcortile.com

Porta Sirena:
Via Ponte Marmoreo
Paestum (SA)
828721035
www.portasirena.it

9. BASILICATA

Carrera della Regina:
Genzano di Lucania
SP 169 km
Genzano di Lucania (PZ)
3497611453
www.carreradellaregina.it

10. CALABRIA

Santa Marina:
contrada Santa Marina
Oriolo (CS)
981931519
www.biosantamarina.it

Dattilo:
Contrada Dattilo
Strongoli (KR)
962865613
www.dattilo.it

Le Puzelle:
SS 107 Località Puzelle
Santa Severina (KR)
www.lepuzelle.it

11. PUGLIA

Torrevecchia:
Contrada Torrevecchia
San Pancrazio Salentino (BR)
3388287360
www.torrevecchia.com

Serragambetta:
Via per Conversano 204
Castellana-Grotte (BA)
804962181
www.serragambetta.com

Narducci:
Via Lecce 144
Speziale di Fassano (BR)
804810185
www.agriturismonarducci.it

12. SICILIA

Santa Mamma:
Via Nazionale
Acquedolci (ME)
3476792228
www.masseriasantamamma
.businesscatalyst.com

Tenuta Roccadia:
Contrada Roccadia
Carlentini (SR)
095990362
www.roccadia.com

Baglio Fontana:
Via Palermo 1
Buseto Palizzolo (TP)
923855000
www.bagliofontana.it

Metric Conversion Tables

Approximate U.S. Metric Equivalents

LIQUID INGREDIENTS

U.S. MEASURES	METRIC	U.S. MEASURES	METRIC
¼ TSP.	1.23 ML	2 TBSP.	29.57 ML
½ TSP.	2.36 ML	3 TBSP.	44.36 ML
¾ TSP.	3.70 ML	¼ CUP	59.15 ML
1 TSP.	4.93 ML	½ CUP	118.30 ML
1¼ TSP.	6.16 ML	1 CUP	236.59 ML
1½ TSP.	7.39 ML	2 CUPS OR 1 PT.	473.18 ML
1¾ TSP.	8.63 ML	3 CUPS	709.77 ML
2 TSP.	9.86 ML	4 CUPS OR 1 QT.	946.36 ML
1 TBSP.	14.79 ML	4 QTS. OR 1 GAL.	3.79 L

DRY INGREDIENTS

U.S. MEASURES		METRIC	U.S. MEASURES	METRIC
17⅗ OZ.	1 LIVRE	500 G	2 OZ.	60 (56.6) G
16 OZ.	1 LB.	454 G	1¾ OZ.	50 G
8⅞ OZ.		250 G	1 OZ.	30 (28.3) G
5¼ OZ.		150 G	⅞ OZ.	25 G
4½ OZ.		125 G	¾ OZ.	21 (21.3) G
4 OZ.		115 (113.2) G	½ OZ.	15 (14.2) G
3½ OZ.		100 G	¼ OZ.	7 (7.1) G
3 OZ.		85 (84.9) G	⅛ OZ.	3½ (3.5) G
2⅘ OZ.		80 G	¹⁄₁₆ OZ.	2 (1.8) G

INDEX

ABOUT THE AUTHORS

Matthew Scialabba and **Melissa Pellegrino,** a husband-and-wife cooking and writing team, met while both living in Italy. Their shared passion for Italian food and culture led them to embark on many culinary journeys throughout the country. These experiences have included apprenticing at a Ligurian vineyard, graduating from a professional culinary school in Florence, and working and cooking together at several Italian agriturismi, which eventually sparked their dream of writing their first book, *The Italian Farmer's Table,* and its sequel, *The Southern Italian Farmer's Table.* For more information about the authors and this project, please visit their website at www.theitalianfarmerstable.com. They live in Guilford, Connecticut and are the owners of Bufalina Wood Fired Pizza.